BLITZKRIEG IN COLOUR

Rare German Photographs from World War II

Text by Simon Forty
Images from Ian Stewart Spring

Greenhill Books

Acknowledgements (from Simon Forty)
Thanks to Ian Spring for his help and his magnificent collection; to Patrick Hook and my collaborator Richard Charlton Taylor for advice, corrections and material help in compiling the captions; and to editor Jessica Cuthbert-Smith and designer Ian Hughes for making it read well and look so good.

Acknowledgements (from Ian Spring)
As I have progressed with further book publications with Greenhill, I have been blessed with fantastic support from my fellow photo collectors and historians around the world. This book has been in many ways a worldwide project to create something very special and new that no one has ever seen before. A collaboration of unique colour photographic material from World War II sourced from the finest private photographic collections in the world. I give special thanks to the following wonderful people:

Fiona Frances Spring (Germany) my amazing daughter for all her help scanning photo slides for daddy!

Michael Leventhal (UK), Simon Forty (UK), David Williams (USA), Matthew Acred (Australia), Akira Takiguchi (Japan), John Avis (USA), Michael Sobotta (Germany), Matthieu Comas (France), Philippe Ricco (France).

One extra note. I wanted to add a massive thank you to the amazing photo slide collector David Williams in the United States for sharing some of his incredible treasures with me. I hope we can continue to make world-class colour photography books into the future: showing the world what our history truly looked like.

Front cover: Panzer-Regiment 5, 3. Panzer-Division took part in heavy fighting at Gembloux on May 15, drawing the Allies into Belgium and masking the main German attack in the south.

Please note that all dates are 1940 unless specified otherwise.

Blitzkrieg in Colour

Greenhill
Books

First published in 2026
by Greenhill Books, Lionel Leventhal Ltd
c/o Pen & Sword Books Ltd,
George House, Unit 12 & 13, Beevor Street,
Off Pontefract Road, Barnsley,
South Yorkshire S71 1HN
www.greenhillbooks.com

CIP data records for this title are available from the British Library

Hardback edition: 978-1-80500-250-5
ePub edition: 978-1-80500-251-2
PDF: 978-1-80500-252-9

Typeset and designed by www.mousematdesign.com
Printed and bound in China by
1010 Printing International Ltd.

The Publisher's authorised representative in the EU for product safety is Authorised Rep Compliance Ltd., Ground Floor, 71 Lower Baggot Street, Dublin D02 P593, Ireland.
www.arccompliance.com

MIX
Paper | Supporting responsible forestry
FSC www.fsc.org FSC® C016973

Contents

Abbreviations and Glossary

Armee	A German army, ie a unit (see also Heer)
Ausf	*Ausführung* = model or mark
BCC	*Bataillon de chars de combat* = tank battalion (French)
BDM	Bund Deutscher Mädel = League of German Girls
BEF	British Expeditionary Force
DCR	*Division cuirassée* = armoured division (French)
DRB	Deutsche Reichsbahn = German Railways
EK 1/2	*Eisernes Kreuz 1./2. Klasse* = Iron Cross 1st/2nd Class
Fallschirmjäger	Paratroops
Feldmütze	Field cap
Gebirgsjäger	Mountain troops
Geschwader	Wing, usually divided into *Gruppen* (groups with Roman numerals) and *Staffeln* (flights with Arabic numerals). There were various types (such as *Kampfgeschwader* = bomber wing, *Sturzkampfgeschwader* = dive-bomber wing or *Jagdgeschwader* = fighter wing. So, II./KG 52 = II. Gruppe Kampfgeschwader 52 and 6./KG 52 = 6. Staffel
Gestapo	Geheime Staatspolizei = Secret Police, part of the SS from 1934
GFM	*Generalfeldmarschall* = field marshal
Heer	The German Army
Heeresgruppe	HG = army group
HJ	Hitlerjugend = Hitler Youth
IR	*Infanterie-regiment* = infantry regiment
Kfz	*Kraftfahrzeug* = motor vehicle
KStN	*Kriegsstärkenachweisungen* = official war establishment (orders of battle of units)
KwK	*Kampfwagenkanone* = armoured fighting vehicle (AFV) gun
leFH	*leichte Feldhaubitze* = light field howitzer
leIG	*leichtes Infanteriegeschütz* = light infantry gun
NKAV	*Nebelkerzenabwurfvorrichtung* = smoke dispenser
NSDAP	Nationalsozialistische Deutsche Arbeiterpartei = National Socialist German Workers' Party – the Nazi Party
OKH/W	Oberkommando des Heeres/der Wehrmacht = High Command of the Heer (Army)/Wehrmacht (Armed Forces)
Pak	*Panzerabwehrkanone* = anti-tank gun
Panzerjäger	Self-propelled (SP) tank destroyer
PzBefWg	*Panzerbefehlswagen* = command tank

PzKpfw	*Panzerkampfwagen* = armoured vehicle, usually tank
RAD	Reichsarbeitsdienst = Reich Labour Service
RK	*Ritterkreuz des Eisernen Kreuzes* = Knight's Cross of the Iron Cross. Higher orders were instituted during the war: *mit Eichenlaub*; *mit Eichenlaub und Schwertern*; *mit Eichenlaub, Schwertern und Diamanten*; *mit Goldene Eichenlaub, Schwertern und Diamanten* – with (golden) oak leaves/swords/diamonds)
RLM	Reichsluftfahrtministerium = German Air Ministry
RM	Reichsmark – German currency
SD	Sicherheitsdienst = SS intelligent agency
SdAh	*Sonderanhänger* = special trailer
SdKfz	*Sonderkraftfahrzeug* = special-purpose motor vehicle
sFH	*schwere Feldhaubitze* = heavy field howitzer
SS	Schutzstaffel. Split into Allgemeine-SS (general), Totenkopfverbände (concentration/extermination camps) and Waffen-SS (armed)
'Ultra'	Shorthand for the decryption of German coded messages by Bletchley Park and other sources
Unternehmen	Operation; used to differentiate between German and Allied operations
Waffenfarbe	Arm of service colours
Westwall	German defensive line along the Belgian and French borders. The Allies called it the Siegfried Line.

Introduction

THE TREATY OF VERSAILLES at the close of the Great War placed the blame for the conflict firmly on the belligerents – Germany and the Central Powers. The terms of the treaty reflected an attempt to stop such a conflict happening again and to make reparations to those who had borne the brunt of the Central Powers' assaults. Key elements affecting Germany were:

- The demilitarisation of the Rhineland.
- Giving France the Saar region – notable for its coalfields – for fifteen years.
- The return to France of Alsace-Lorraine, which had been occupied by Germany in 1871.
- Giving Poland a corridor of land to the west of East Prussia and making Danzig a free city controlled by the League of Nations.
- Giving France and Britain all Germany's colonies.
- Forbidding Germany to unite with Austria.
- Restricting the size of German armed forces (army to 100,000 men), with many other strictures (no submarines; no air force).
- Germany would have to pay reparations (later set at 132 billion gold marks).

These terms are often said to have been at the root of Second World War II and certainly few people were satisfied with the outcome. The British prime minister, David Lloyd George, thought the terms too harsh and is reported as saying, 'We shall have to fight another war in 25 years' time.' US President Woodrow Wilson also felt them too stringent, as did the US Senate, which refused to sign the treaty. The Americans would help prop up the Weimar Republic – particularly through the 1924 Dawes Plan and 1929 Young Plan – until the financial crisis of the late 1920s and early 1930s upended the world economy.

Germany, of course, hated every aspect of the treaty – having to accept guilt for having started the war, their loss of territories, the reparations – and the Weimar Republic's acceptance of it made the government an easy target for the NSDAP. Hitler's rhetoric railed against the treaty and the government which had signed it, and he promoted the idea that the German Army hadn't lost the war but had been stabbed in the back by a coterie of Jewish financiers, Bolsheviks and government criminals.

His arguments took the NSDAP to power and at the centre of the Nazis' agenda was repudiation of the treaty, refusal to pay reparations and the return of its lost territories.

Everyone should have known what was coming. Hitler had identified Germany's major enemies in *Mein Kampf* – there were a lot of them, including his odd concoction of Jews and Bolsheviks, of course, but also France.

Above: On 13 March 1930 SA-Stabschef Otto Wegener and Hitler were travelling in a Daimler-built Mercedes 15/70/100 when a lorry crashed into them. They escaped relatively unscathed and Hitler said later, 'It was then I decided to use only a Mercedes for the rest of my life.' From 1931 he used a Series I 770 (W07) – the Grosser Mercedes limousine; from 1938 the Series II 770 (W150). Hitler's version had extensive armour-plating and 40mm (1.6-inch) glass.

In this 1938 photo, Hitler acknowledges the adoring Berliners while his SS bodyguard and valet sit behind him. His primary chauffeur from 1936 to his death was SS member Erich Kempka, who delivered the petrol used to burn his and Eva Braun's bodies after they committed suicide.

Right: The young teenage element of the Hitlerjugend organisation was the Deutsches Jungvolk (DJ – German Youth). Here their fanfare trumpets are adorned with silver bullion-edged, double-sided, black linen banners with a white/silver Sig-rune stitched in the centre. The rune was the first official flag adopted as the emblem of the DJ.

Above: The 1930s saw militaristic marches and rallies throughout Germany – this one is in Lübeck on the Baltic – to project the NSDAP and a strong and united image of Germany. They were carefully orchestrated to reinforce the idea of a German community (*Volksgemeinschaft*), the myth of the Führer and to remember the honoured dead. They sometimes lasted for several days and involved military manoeuvres, mass parades, speeches, propaganda exhibitions, displays of war booty and other events, latterly to evoke support for the war effort.

Right: Accordions were invented in Germany and much used for wartime entertainment. These players are starting young: 6–10-year-olds were the Pimpfen – cubs; 10–14 the Deutsches Jungvolk; thereafter they joined the Hitlerjugend. There were few absentees: if you didn't join, many doors closed, not only in youth but in later life – from 1936, the Hitlerjugend was the only organisation for boys, and any rival groups were banned. By 1938 it had 7.7 million members.

> We must be quite clear on the following point: France is and will remain the implacable enemy of Germany. It does not matter what governments have ruled or will rule in France, their foreign policy will always be directed towards acquiring possession of the Rhine frontier and consolidating France's position on this river by disuniting and dismembering Germany.

Hitler had also identified why he had to attack Poland: *Lebensraum.*

> … we are not fighting for our position as a world power but only for the existence of our country, for national unity and the daily bread of our children. … [We] must stick firmly to the aim that we have set for our foreign policy; namely, that the German people must be assured the territorial area which is necessary for it to exist on this earth.

He had lined up his excuses:

> I must protest as sharply as possible against those nationalist scribes who pretend that such territorial extension would be a 'violation of the sacred rights of man' … For no nation on earth possesses a square yard of ground and soil by decree of a higher will and in virtue of a higher right. The German frontiers are the outcome of chance, and are only temporary frontiers that have been established as the result of political struggles which took place at various times. … State frontiers are established by human beings and may be changed by human beings. … If the German people are imprisoned within an impossible territorial area and for that reason are face to face with a miserable future, this is not by the command of destiny, and the refusal to accept such a situation is by no means a violation of destiny's laws. … It can be solved only by the winning of such territory for the settlement of our people as will extend the area of the motherland and thereby will not only keep the new settlers in the closest communion with the land of their origin, but will guarantee to this territorial ensemble the advantages which arise from the fact that in their expansion over greater territory the people remain united as a political unit.

His excuses included a pre-emptive pardon for himself and senior Nazis for leading the country into war:

> The territory on which one day our German peasants will be able to bring forth and nourish their sturdy sons will justify the blood of the sons of the peasants that has to be shed today. And the statesmen who will have decreed this sacrifice may be persecuted by their contemporaries, but posterity will absolve them from all guilt for having demanded this offering from their people.

A distinctive insignia helps spark interest in a political movement, as Hitler noted. The black Nazi swastika on a white circle in the middle of a red flag was both noticeable and powerful, its colours harking back to the black and white of the Prussian flag. The photo captures the moment a motorcade works its way around the rally. The group in the centre between the banners are band members from the Reichsarbeitsdienst (RAD – Reich Labour Service), the paramilitary organisation that took all young men between the ages of 18 and 25 into six months of hard labour before their military service.

While the Nazis said that women were best keeping the home and bearing children, they actually used them from the start of the war in roles that supported the Wehrmacht. The Wehrmachtshelferinnen (female Wehrmacht auxiliaries/helpers) were mainly recruited from the female equivalent of the Hitlerjugend, the Bund Deutscher Mädel (BDM – League of German Girls). Involvement in the BDM was mandatory from 1939 for girls aged 14 to 18 (the younger 10–14-year olds joined the Jungmädelbund – Young Girls' League). As with the boys, sport was intended to create strong and healthy young people. The photo shows two BDM using a *Rhönrad*, a gym wheel, invented by a German metalworker in 1925 as a means of building full-body strength and stamina.

A German air force had been expressly forbidden at Versailles, but clandestine development – often linked into civil aircraft – including the agreements with the Soviet Union, meant that by the late 1930s the Luftwaffe was probably the best military air force in the world. Its numbers weren't as large as Hitler had hoped, but it performed brilliantly in support of the army over Poland and the west. No Nazi rally or march could be complete without flights of modern aircraft to show the country's military might: these are Messerschmitt Bf 110s over a Berlin rally in 1938.

The Italian foreign minister, Galeazzo Ciano, Mussolini's son-in-law, watches a sports day in Germany in 1939. Behind him is the Fascist Italian flag and the cog and swastika of the Deutsche Arbeitsfront (German Labour Front). Italy's relationship with Germany was complicated. Hitler needed allies in the run-up to war, but the Italian economy was never able to supply its army with the tools to do the job, leading to the Germans having to bail out the Italians after their ill-advised attacks on Greece and the British in Africa. Ciano was against Italian involvement in the war and became a critic of his father-in-law, who eventually had him executed on 11 January 1944, along with other members of the Grand Council of Fascism who had deposed Mussolini after the Allied invasion of Sicily in favour of King Victor Emmanuel III.

The Hall of Liberation (Befreiungshalle) on Mount Michelsberg outside Kelheim was commissioned by King Ludwig I of Bavaria to commemorate the victories over Napoleon in 1813–15. Its symbolic association with German nationalism and triumph over a foreign power made it an important place for the Nazis. The words in the marble flooring struck a chord: 'May the Germans never forget what made necessary the struggle for freedom and by what means they won.' On 22 October 1933 Hitler spoke there and then watched a *Sturmabteilung* parade.

These 'Brownshirts' played a significant role in Hitler's rise to power, using intimidation to disrupt opposition meetings and marches. However, they became too much of a liability and were purged and their leaders killed in the 1934 Night of the Long Knives. Note the *Alte Kämpfer* – old fighters – on the far bus. These were the people who joined the party before the 1930 federal election and some, the Blood Order, had taken part in the 1923 Munich Beer Hall Putsch.

An anxious *SA-Truppführer* in greatcoat and kepi with emerald-green collar tabs awaits a visitor. His men wear an eclectic mixture of uniforms. The SA's distinctive *Dienstmütze* (or *Schaftmütze*) kepi – nicknamed the 'coffee-can hat' – came in different colours, denoting the *Gruppe* (division), these with black crowns identify Berlin.

Most countries in Europe in the 1930s had their own fascist parties. Britain had Mosley and the British Union of Fascists; France had the Parti populaire français (PPF – French Popular Party) led by Jacques Doriot. Formed in 1936 and financed in part by the Italian Fascists, after France fell the PPF was supported by the Nazis and collaborated with the deportation of Jews. Doriot served in the Légion des volontaires français contre le bolchévisme and had three tours of duty on the Eastern Front. He died on 22 February 1945 when his car was strafed by an Allied aircraft.

The Soviet Agreements

On 3 March 1918, the Soviet Union was forced to sign the Treaty of Brest-Litovsk with the Central Powers. Its draconian terms make those of Versailles seem benevolent. It lost Finland, Ukraine, its Polish, Lithuanian and Baltic territories, parts of Belo-Russia, Bessarabia and areas in the Caucasus. These territories accounted for 34 per cent of the tsarist Russian Empire's population, a third of its richest agricultural areas, nearly half of its European territory and 54 per cent of its industrial areas. It also agreed to pay 6 billion marks in reparations.

In fact, the Armistice of 11 November 1918 saw the Treaty of Brest-Litovsk abrogated, but that didn't stop the Civil War that continued on to October 1922 and included the involvement of Britain, the United States, France, Czechoslovakia and Japan. By the end, the Soviets had regained most of Ukraine, its eastern territories and some of its territory in the Caucasus, although independence movements would continue the fighting into the 1930s.

In the 1920s, in order to circumvent the Treaty of Versailles which expressly forbade the Germans from having an air force and the army from having tanks, the Germans established more cordial arrangements with the Soviet Union. The subsequent Treaty of Rapallo in April 1922 saw the two countries 'mutually agree to waive their claims for compensation for expenditure incurred on account of the war, and also for war damages', and 'The two Governments shall co-operate in a spirit of mutual goodwill in meeting the economic needs of both countries.' It also had a clandestine side which led to the Germans setting up a number of bases in the Soviet Union. There was an airfield at Lipetsk north of Voronezh from 1924; by 1933 over 1,200 Luftwaffe pilots had trained there. 'Tomka' was the codename for a gas warfare school for training and technology at Podosinky near Saratov set up in 1926. Perhaps the most important was 'Kama', a *Panzerschule* – combat vehicle school – at Kazan from 1926. The tanks – six 23-ton vehicles with 75mm main armament and three 12-ton tanks with 37mm guns – arrived in 1929. Amongst the teachers was Heinz Guderian. There were Soviet observers in Germany, and German manufacturers were also in action in the Soviet Union.

This friendly approach ended when Hitler came to power. He and the Nazis were profoundly anti-Bolshevik, so it came as a major surprise when Germany and the Soviet Union made two agreements in August 1939. The first, on the 19th, was a trade and credit agreement, worth 200 million RM over seven years, for the Soviets to buy capital goods (mainly factory equipment, tools and various means of transport). The second, often called the Molotov–Ribbentrop Pact, was on 23 August 1939. The *Manchester Guardian* reported on the 22nd:

> Germany and the Soviet Union have agreed to conclude a pact of non-aggression.
>
> This surprising announcement was made in Berlin last night by the official German news agency. … The news was completely unexpected.
>
> The decision to sign the pact is announced only the day after the signing of a commercial treaty was made known. The Russian press yesterday, suddenly

> changing its tone towards Germany, warmly welcomed the commercial agreement as being likely to lead to better political relations.

On the 25th it said:

> It is more and more evident that the Russo-German Pact has far greater significance than the published terms imply, though even they would be significant enough. There is a growing belief that, as was suggested in these columns yesterday, Germany and Russia have agreed on the partition of Poland and that the Baltic States are to be a Russian sphere of influence.

The agreement freed Stalin to attack Japan in the Far East without fear of attack in the west. It freed Germany to attack Poland – with all the possible problems that might cause with France and Britain – without fear of attack by the Soviet Union. Indeed, the secret protocols went further – as the *Manchester Guardian* had suggested – and agreed the division of Poland and that the Baltic states would be within the Soviet zone.

Another agreement was signed on 11 February 1940. This commercial arrangement saw the Soviets agree to deliver to Germany c.420 million RM worth of fuel, raw materials and food, and to facilitate the purchase of goods that the British blockade was restricting. Ironically, many of the commodities that the Soviet Union supplied went towards the forces that were used to invade the Soviet Union in 1941.

It is interesting to note what Hitler said about alliances in *Mein Kampf* and, in particular, a Russo-German alliance:

> An alliance which is not for the purpose of waging war has no meaning and no value. Even though at the moment when an alliance is concluded the prospect of war is a distant one, still the idea of the situation developing towards war is the profound reason for entering into an alliance. It is out of the question to think that the other powers would be deceived as to the purpose of such an alliance. A Russo-German coalition would remain either a matter of so much paper – and in this case it would have no meaning for us – or the letter of the treaty would be put into practice visibly, and in that case the rest of the world would be warned. It would be childish to think that in such circumstances England and France would wait for ten years to give the Russo-German alliance time to complete its technical preparations. No. The storm would break over Germany immediately.
>
> Therefore, the fact of forming an alliance with Russia would be the signal for a new war. And the result of that would be the end of Germany.

Mein Kampf, it has to be said, is a little like the Bible and cherry-picking can excuse most actions – but at least he got one thing right! The new war was the end for Nazi Germany.

What he didn't take into account in *Mein Kampf*, written in 1925, was that in 1939, he and his country would be buoyed up by six years of Nazi rule that had seen the Rhineland reoccupied by German troops in repudiation of the Treaty of Versailles, Austria incorporated into the Reich through the Anschluss, and the Sudetenland – part of the then country of Czechoslovakia – annexed.

These political successes had walked hand in hand with the rebuilding of the German armed forces – the Wehrmacht – the militarisation and radicalisation of the country, especially the youth, and a massive rearmament campaign. In 1934, Hitler had told a trusted coterie of SS, SA and Wehrmacht that Germany would be ready for war in eight years. In 1936, his four-year plan memorandum had cut two years from this: Germany had to be ready for war in 1940. In 1939, however, the planets aligned and Hitler decided that there was nothing to be gained by waiting. He had quoted Schiller's *Wallenstein* in *Mein Kampf*:

> *Und setzet ihr nicht das Leben ein,*
> *Nie wird euch das Leben gewonnen sein.*
> (And if you do not stake your life,
> You will never win life for yourself.)

On 31 August 1939, as Supreme Commander of the Armed Forces, he promulgated the first of his war directives. On 1 September 1939, Germany invaded Poland. The Blitzkrieg had begun.

Mounted NCOs on well-groomed horses lead an infantry column past welcoming crowds. The location is in Alsace-Lorraine, but the people could be ethnic Germans in Poland or central Europe. Ethnic Germans were settled in Czechoslovakia, Poland, Hungary, Yugoslavia, Romania, Estonia and along the Baltic, and most welcomed the Germans when they arrived. During the early months of the war, it wasn't uncommon for the military to be presented with flowers, and many have orange blooms on their tunics. In 1945, however, the Germans who had remained in or had been transplanted into captured territories faced terrible retribution. The lucky ones were the 14 million (the West German government figure) who were expelled – many from Poland and Czechoslovakia. Some figures put the deaths as high as a million.

Condor Legion

Right: The German involvement in the Spanish Civil War between 1936 and 1939 on the side of Generalissimo Franco's Nationalists was substantial. Providing military aid, personnel and equipment – particularly tanks, aircraft and anti-tank weapons – against the democratically elected Soviet-backed Republicans loyal to the left-leaning government, the Condor Legion was a combination of Luftwaffe and Heer troops. There weren't many combat troops and the Germans were mainly used in a training role. It was the Luftwaffe that supplied many of the transport aircraft to bring the Spanish Army of Africa to Franco's aid in July 1936. The photo shows a group of German pilots and aircrew, one of whom holds the standard of the Condor Legion, under an early Heinkel He 111E. The Spanish Civil War was a useful testing ground for German weapons and tactics. (Akira Takiguchi)

Right: The bright-red collar tabs with three gulls identifies the wearer as an *Obergefreiter* (private first class) in a Luftwaffe anti-aircraft unit. On his right breast he wears the *Spanienkreuz* – Spanish Cross – in silver with swords. Over 8,000 were awarded to fighting men of the Wehrmacht who took part in the Spanish Civil War as members of the Condor Legion. This *Obergefreiter* is enjoying a break on the beach at Swinemünde, a spa city on the Baltic, today Świnoujście in Poland. He's standing next to a wicker *Strandkorb*, a traditional hooded beach chair windbreak that could seat two.

Below: The calm before the storm. Berlin is bright and cheerful on 6 June 1939, as it prepares for the Condor Legion to be honoured in a parade as part of Hitler's fiftieth birthday celebrations.

Directive No. 1 For the Conduct of the War

1. All political means having been exhausted to correct in a peaceful manner the unbearable situation on Germany's eastern border, I have decided upon a solution by force.

2. The attack against Poland is to be carried out according to preparations for *Fall Weiss*, with the modifications resulting from the Army's having been almost completely deployed in the meantime.

The assignment of missions and the operational objective remain unchanged.

Date of attack: 1 Sep. 1939

Time of attack: 04:45 [according to the War Diary of the German Naval Staff]

This time will also apply with regard to the operations at Danzig and the Gulf of Danzig, and at the Dirschau bridge.

3. In the West it is important to leave the responsibility for opening hostilities unmistakably to Britain and France. Minor border violations are for the time being to be countered with purely local actions.

The assurances of neutrality given by us to the Netherlands, Belgium, Luxemburg and Switzerland are to be meticulously observed.

The western border of Germany must not be crossed by land at any place, without my explicit authorization.

The same applies at sea for all hostile actions or any which can be so construed.

The defensive measures of the Air Force are, for the time being, to be limited entirely to repulsing enemy air attacks at the Reich border. The borders of neutral states are to be respected as long as possible in warding off single aircraft or small formations. Only when air protection is no longer assured in the West on account of the employment against German territory of stronger French and British attacking forces over neutral countries, is defense over neutral territory to be permitted also.

It is particularly important that any violation of the neutrality of these countries by the enemy in the West should be speedily reported to the Armed Forces High Command.

4. Should Britain and France begin hostilities against Germany, it will be the duty of the armed forces operating in the West to ensure the victorious conclusion of the operations against Poland, although conserving their forces as much as possible. Within these limits, the enemy forces and their economic resources are to be damaged as much as possible. I reserve for myself the right to order offensive actions in all cases.

The Army will hold the Westwall, and will be prepared to prevent the Western Powers from flanking it in the north, through violation of Belgian or Dutch territory.

If French forces move into Luxemburg, the bridges at the frontier may be blown up.

The Navy will carry out the war on merchant shipping with Britain as the focal point. For intensification of the effect, the declaration of danger zones can be expected. The Naval High Command is to report in what waters and to what degree the establishment of danger zones is considered advisable. The text of a public declaration is to be prepared in collaboration with the Foreign Office, and is to be submitted for my approval through the Armed Forces High Command.

The Baltic Sea is to be protected against enemy penetration. The Commander-in-Chief, Navy will decide whether the entrances to the Baltic may be mined for this purpose.

The Air Force is, first of all, to prevent the employment of the French and British air forces against the German Army and the German lebensraum.

In the strategy against Britain, preparations are to be made for the Air Force to disrupt Britain's imports, her armament industry and her troop-transports to France. Any favourable opportunity for an effective attack against massed vessels of the British fleet, especially battleships and aircraft carriers, is to be utilised. The decision regarding attacks on London rests with me.

Attacks on British home territory are to be prepared, bearing in mind that inconclusive results with insufficient forces are to be avoided under all circumstances.

(All Führer Directive quotes are from OSS, 1948)

The Anschluss

Hitler's clever political brinkmanship saw Germany grow in size and might during the latter half of the 1930s. Expressly forbidden at Versailles, the Anschluss of 12 March 1938 saw Austria annexed. Germany laid hands on that country's Central Bank's reserves, natural resources (primarily iron ore) and its population, over a million of whom would serve in the Wehrmacht and Waffen-SS. This atmospheric double exposure shows how the crowds packed together for a glimpse of the Führer.

Austrian troops march under Nazi banners. The Austrian Army had a mobilisation level of approximately 127,000, of which 60,000 were standing. They were integrated into the Wehrmacht and, as with Poland and the Sudetenland, their territory was divided into Wehrkreise, XVII in Vienna and XVIII in Salzburg. Conscription was reintroduced. In total 950,000 Austrians fought for the Wehrmacht; a further 150,000 for the Waffen-SS.

The German Army learned a great deal about its logistics systems and mobility during the movement into Austria and Czechoslovakia. Mechanical deficiencies, breakdowns, lack of fuel, poor training – all of these factors would dog the Wehrmacht and Blitzkrieg. This was one reason why the German Army insisted that they needed time after *Fall Weiss* to learn lessons, integrate new troops and, above all, train all arms before the much greater operations in the west.

1 The Wehrmacht in 1939

THE MECHANISATION OF WARFARE that took place in the interwar years and was unleashed against Poland in 1939 led western journalists to popularise the term that most catchily summed up the seemingly unstoppable Nazi juggernaut: *Blitzkrieg* – lightning war.

Compared to the trench warfare so well remembered from World War I, the deep penetrations led by armoured columns supported closely from the air – particularly Ju 87 Stuka dive-bombers – seemed revolutionary. By the time the dust had settled in June 1940, Blitzkrieg had seen the political map change to an overwhelming degree; German troops dominated the continent from the English Channel to the new Soviet border in Poland and from the Arctic to the Mediterranean. Further devastating assaults would condemn the Balkans, Greece and Crete to long years of subjugation by a victor whose military successes allowed their hardline ideology carte blanche in the conquered territories.

However, while the historians of the immediate postwar period tried to argue that Blitzkrieg was a carefully analysed strategy that was based on the interwar ruminations of British and German theorists, recent scholarship suggests it was nothing of the sort. It is true that German tactics were different to those of World War I – how could they not be? Technological developments to aircraft, tanks and military vehicles in general had changed the face of war. The British forces, for example, when they arrived in France in 1939, were completely motorised. But in September 1939, the idea of 'Blitzkrieg' didn't exist.

What did exist was a finely tuned army that was raring to put its training into practice. The troops and their leaders were highly motivated with high morale. Leadership was the critical factor in what we think of as Blitzkrieg. This ensured all-arms cooperation, flexibility of operations and continuity of command even when the officers were incapacitated. German leadership in battle was based on *Auftragstaktik* – the mission-based tactical approach that was so much a part of German NCO and young officer training. The bible for this approach was 'Tante Frieda' – *Truppenfuhrung*, the 1933 update of the earlier *Fuhrung und Gefecht der verbundenen Waffen* (Command and Combat Employment of Combined Arms) produced in 1921.

> Mission command consisted of the issuing of short operational orders which gave subordinates a concise and clear idea of the commander's notion of the mission and a list of coordinating measures, such as unit boundaries. The order also provided all the support assets available to the higher commander to facilitate the subordinate's accomplishment of the mission. How the subordinate accomplished the mission was up to his own devices. (McGrath, 2013)

GFM Erich von Manstein thought, 'Individual leadership was fostered on a scale unrivalled in any other army right down to the most junior NCO or infantryman and in this lay the secret of our success.' It was certainly true that the detailed training given to troops after the repudiation of the Treaty of Versailles meant that in the early war years there were plenty of examples of *Auftragstaktik* in action. One such was the capture of the fort at Eben Emael. After the commander of the operation, Leutnant Rudolf Witzig, was forced to make an emergency landing shortly after take-off, the senior NCO, Hauptfeldwebel Helmut Wenzel, took command and continued the mission until Witzig was able to relaunch. Both received the *Ritterkreuz* (RK). It was this flexibility that meant that, even late in the war, German units could be combined together in ad hoc *Kampfgruppen* and perform without friction.

However, while the attack on Poland led to a swift conclusion, German combined-arms tactics weren't battle proven when it started. True, the armour wasn't dissipated into penny packets as infantry support – as it was in most armies of the time – but *Fall Weiss* wasn't spearheaded by Panzers in the same way that *Fall Gelb* would be. After the event, from October 1939 the Germans subjected the Polish campaign to a searingly honest appreciation of what had happened in order to improve training and tactics. As Williamson Murray put it in an article in *Armed Forces & Society*: 'This willingness to be self-critical was one of the major factors that enabled the German army to perform at such a high level throughout World War II.' As an example of this, analysis of the performance of the *leichten Divisionen* (light divisions) led to their becoming full-blown Panzer divisions. In their turn, the Panzer divisions were criticised for being too Panzer heavy, lacking infantry, and the infantry–tank cooperation hadn't worked as well as it should. The Panzer losses – nearly 700 of them, of which 217 were written off – were quite high and it was the infantry that received most of the plaudits. The PzKpfw Is and IIs had proved susceptible to Polish anti-tank rifles. Nevertheless, it was the speed of the German attack – significantly helped by the Panzers – that was the main reason for the Polish Army's collapse. Unrelenting, it gave the Poles no chance to construct a suitable defensive line.

The main result of the analysis of the Wehrmacht's actions in the fighting was a training emphasis on 'reconnaissance and security, march discipline, infantry fire discipline, especially after long marches, cooperation between weapons, offensive and defensive tactics at twilight and during the night, and conversion from an offensive to a defensive posture' (Murray, *Armed Forces & Society*, 1981). In particular, the training was aimed at the many new recruits, reservists and Landwehr soldiers (older reservists) who needed to be brought up to scratch before the attack on the west. The speedy success of *Fall Gelb* saw the combined-arms tactics blossom and the true power of the Panzer-led army. The Allied armies could do nothing to withstand an assault they knew was coming.

In all the discussions about new weapons and new tactics, it must never be forgotten what the German war aims were and what resulted. As already discussed, Hitler wanted *Lebensraum*, room for the German people to become autarkic, self-sufficient. Those who lived in the areas he wanted to annex would either be evicted or starved to death, their food sent to the Reich.

Hitler's rise to power had mortgaged German industry and the country's wealth – what was left of it after the first war, the reparations and the financial issues of the 1920s – to such an extent that it needed the loot from its conquests to paper over the huge economic issues

A new recruit. He's carrying his *Bekleidungssack* M31 in canvas with leather trim. It contains everything that doesn't fit inside his *Tornister* (rucksack) including, quite possibly, a camera. His camouflaged *Zeltbahn* tent quarter is rolled; his helmet hangs at the back. Had he not been wearing his *Mantel* (overcoat) with its smart dark-green felt collar and shoulder straps, it would be folded and strapped alongside the *Zeltbahn*. His felt-covered *Feldflasche* (water bottle) with its black – probably aluminium – *Trinkbecher* (cup) obscures his *Brotbeutel* (bread bag) on his right hip; his bolt-action Kar98k – Karabiner 98 kurz – is at his shoulder, the edge of his M33 *Patronentaschen* (ammunition pouches) just visible on his *Koppel* (belt) that has a standard *Koppelschloss* (belt buckle). This is adorned with swastika and eagle, above which run the words *Gott Mit Uns* – God with us – rather ironic considering the Nazis' views on religion. The touch of red on his *Feldmütze* indicates he's an artilleryman.

The quality of the German Army's combat engineers (*Pioniere*) was one of the reasons for the speed of movement that epitomised Blitzkrieg. Their many tasks included mine clearance, demolishing strongpoints, clearing barbed wire and overcoming obstacles in support of combat units. The *Sturmpioniere* were trained for assault operations, including close-quarter combat. Note their black *Waffenfarbe*. Standing in the centre of the photo is Feldwebel (Sergeant) Eugen Kastner of 212. Infanterie-Division's IR 316. A number of the men carry pick helves for an axe, mattock or a pickaxe, and an excellent weapon in its own right. The cycles and horses suggest that this group and the others might be a recce group out training; Kastner's lack of medal ribbons and the very new helves suggests this is an early war photo.

that beset the regime. The result of Nazi conquest was rape, slaughter and pillaging as the countries they conquered were fleeced of assets – cultural as well as financial. Their populations condemned to what was, in effect, slavery to help German industry accommodate the mobilization of its armed forces, and the slaughter of those the Nazi state had deemed undesirable: primarily Jews, but also a long list of other unfortunates for whom conquest meant being murdered or worked to death.

The funnelling of huge resources into rearmament saw a significant commitment to the Luftwaffe and Kriegsmarine – not just U-boats but also capital ships, major users of steel. This meant that the spearhead of Blitzkrieg – the nascent Panzerwaffe – perhaps didn't receive the levels of funding it needed, although it's unsurprising that pocket battleships to tackle British naval dominance seemed strategically more important than tanks, which were a new and untested arm.

Panzer development had started in the 1920s as the German Army began tests in the Soviet Union, creating a small training school to examine automotive mechanics – but, as Heinz Guderian detailed in a postwar report:

> ... from 1928 on the curriculum included tactics, the combat employment of tanks, their cooperation with other arms, and their defense. ... In 1930 the first Panzer, antitank, reconnaissance, and motorcycle companies came into existence, though still equipped with dummies. ... Numerous technical tests were made, especially as to cross-country mobility and radio communication, and in 1931 the first units ... appeared at the maneuvers as combat troops.

The first large-scale manoeuvres took place in summer 1935 and shortly after that the first three Panzer divisions were formed. By late 1936 the first units were seeing action in the Spanish Civil War as Oberst Wilhelm Ritter von Thoma's command included over a hundred PzKpfw Is.

However, while the German propaganda machinery showed Panzers and halftracks and a motorised army at every opportunity, the reality was a long way from that. Most of the halftracks were prime movers for artillery, or were used as weapon platforms, specialist vehicles or engineer vehicles. The main personnel carriers were the SdKfz 250 and 251 (c.25,000 in total). Guderian again:

> Originally the German panzer division consisted of two panzer regiments of two battalions each, one infantry brigade having one regiment of two battalions, and a motorcycle battalion. The infantry regiments were to be equipped entirely with SPWs [*Schützenpanzerwagen* = armoured personnel carrier], but up to 1940 only one regiment of 1. Panzerdivision was equipped according to this plan.

This reality needs to be borne in mind whenever one studies KStNs or any of the 'perfect' outlines of what components are supposed to make up a unit. Added to this, as well as paucity of material, the length of time it took to re-equip – almost always involving retraining –

The photographic coverage of the actions of the German Army – particularly in the early war years – was well recorded by the *Propaganda-Kompanien*. The *PK-Berichter* and their equipment have been well covered in earlier titles in this series; suffice it to say here that their proximity to the front and their use of colour photography meant magazines such as *Signal* and *Die Wehrmacht* provided a striking commentary on the battles. This *PK-Berichter* is Hans Wagner, wielding his Leica G. His helmet is the M35 *Stahlhelm* with a rolled edge and removable side vents. The Wehrmacht shield was eventually removed, by order, as it was considered an attractive target for snipers.

meant that part of a unit could conform to the latest KStN and part to the older version.

Alongside the Panzers there were two important developments that enabled Blitzkrieg. First, communications: German tanks carried radios and were therefore able to follow the tactical orders of the various command levels – from troop leader to regiment or battalion, divisions, corps, armies and army groups – as well as communicating with the second key component of Blitzkrieg, the Luftwaffe.

The other major development was the Panzer engineers, an essential component of armoured warfare. They had to ensure that bridges were able to carry increasingly heavy AFVs – or erect suitable replacements – and to provide assistance to enable the Panzer forces to negotiate ditches and barriers, steep inclines, forests and marshes, and especially minefields.

The 1937–38 engineer battalion KStNs (703, 714, 733, 737 and 742) show that, if everything had been perfect, the Panzer divisions would have fielded a motorised combat engineer battalion composed of a light combat engineer company (made up of a company HQ, three engineer platoons, each with an HQ section and three engineer sections), a light engineer column (HQ and two platoons with eight trucks, one having seven cargo trucks and a Kfz 31 ambulance) and two bridging columns: Types B and K. The former was capable of constructing 83m (272ft) 8-ton and 54m (177ft) 16-ton bridges and various sizes of ferry; the latter was a ready-made box-girder 16-ton bridge of nearly 80m (262ft) length that could carry vehicles up to 20 tons on pontoons or could be used in conjunction with the B Type on land.

Other than the spearheading Panzer forces, the weapons of the German Army of 1939 were, in the main, the same as they had been in 1918: horse-drawn artillery and foot-slogging infantry divisions trying desperately to catch up with the tanks to be able to complete the encirclements that allowed what the Germans called the *Kessel* – the cauldron – to form and then be reduced in a battle of annihilation (*Vernichtungsschlacht*). Von Manstein mentions this in *Lost Victories*:

> The unrealistic impression of the campaign was promoted by the German propaganda effort … The German forces were said to be highly motorised, with many more Panzers in support than were actually available, and with fleets of highly efficient ground-support aircraft available on short notice. Little mention was made of the horse-drawn supply columns, of the infantry divisions which often marched on foot at the rate of 30 miles per day, or of the repeated Luftwaffe bombings of advance German units.

In fact, the German Army depended on horses to drag most of its equipment, artillery and supplies. In 1939 the KStN of a first-wave infantry division provided for over 4,800 horses and everything that went with them: the logistics needed to supply fodder, farriers, veterinarians, etc.

Above: Medals are important to soldiers, showing their comrades that they can cut the mustard. The man at right, whose black collar, *Tresse* (silver filigree around his collar) and shoulderboards show him to be a *Pionier Feldwebel*, has just been awarded the EK 2 (Iron Cross 2nd Class) for bravery. On the day of issue this was worn visibly, thereafter as a ribbon from the second buttonhole – as the man on the left. The Feldwebel also wears the SA Sports Badge first issue in bronze and an SA Brunswick Rally Badge of 1931, the third highest national award of the NSDAP, the Nazi Party. He's talking to an RAD *Feldmeister* (lieutenant). Members of the RAD were often drafted into the army as auxiliary units. These men were attached to 11. Panzer-Division, which first saw action during the invasion of Yugoslavia in 1941.

Right: The collar patches identify this man as a Luftwaffe flight crew *Oberleutnant* (first lieutenant). He has an EK 2 ribbon, an EK 1 on his left breast pocket, a Frontflugspanne für Aufklärer (Front Flying Clasp for Reconnaissance) above the pocket and a Fluqzeugführer- und Beobachterabzeichen Beobachterabzeichen (Pilot/Observer Badge) below. His unit was Aufklärungsgruppe 13 that flew Hs 126 tactical reconnaissance aircraft during the battle of France.

German propaganda promoted the idea of a mechanised army as evinced by this (above) 1. Gebirgsjäger-Division SdKfz 7 towing a 15cm sFH 18 heavy howitzer under the watchful eyes of a motorcycle traffic officer. In fact, the German Army of 1940 was a horse-drawn army as it still was in 1945, although most of the time the infantry marched, as here (below). Grabbing rest if they can, these soldiers at least aren't having to carry their packs, which are loaded onto the baggage train – horse-drawn carts.

Opposite: This artillery despatch-rider attached to 110. Infanterie-Division is all togged-up and kick-starting his pressed-steel-framed BMW R11. The distinctive M34 *Kradmantel* – a vulcanised/rubberised waterproof coat – gave him protection against the elements. On his chest he carries his respirator in its cylindrical metal container and a gas cape in its protective carrier. Slung across his back is a Kar98k carbine, the standard infantry rifle. Unusually, he has a Perspex/plexiglass shield attached to his helmet to protect his face, something not seen in many contemporary photos. (Akira Takiguchi)

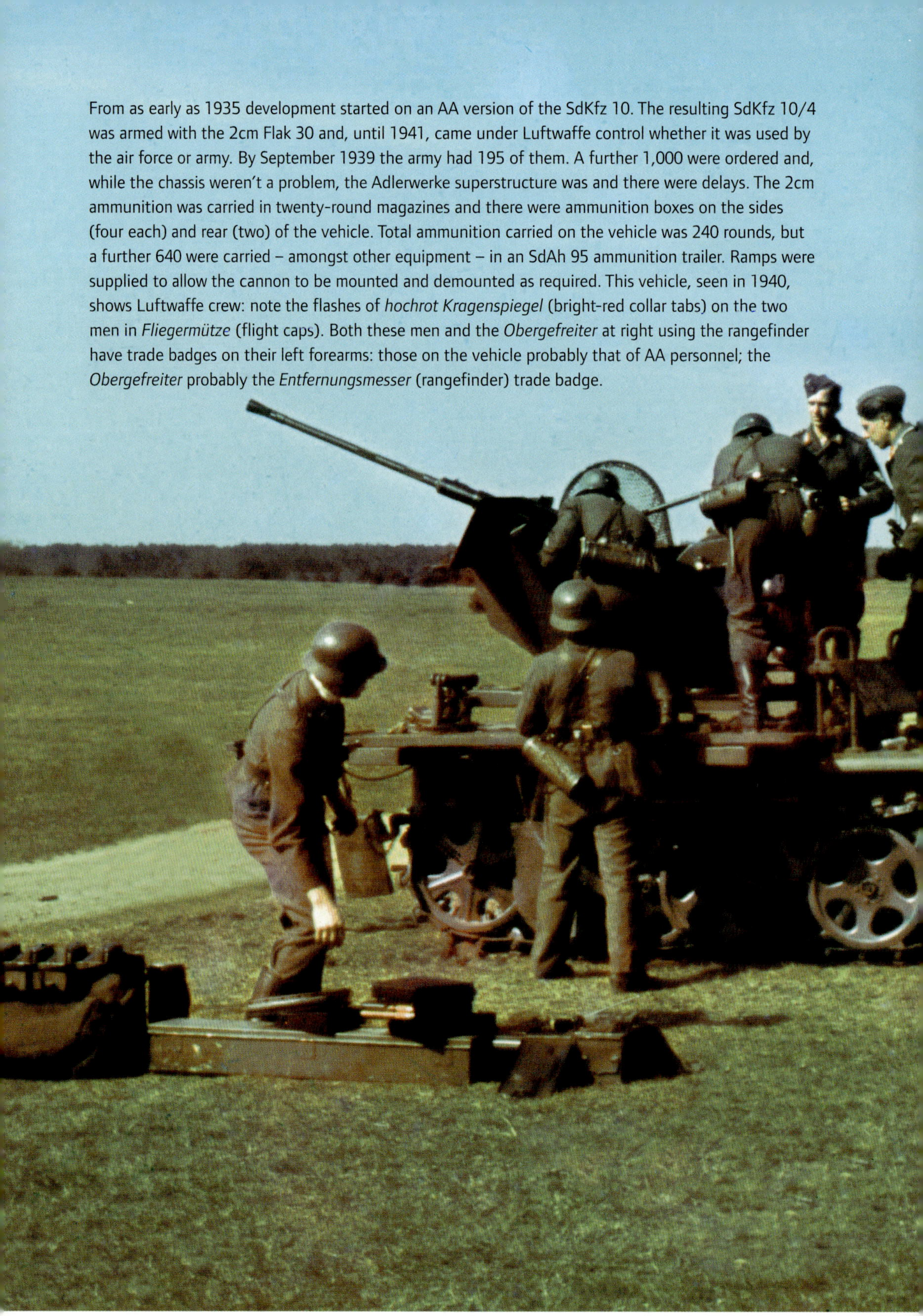

From as early as 1935 development started on an AA version of the SdKfz 10. The resulting SdKfz 10/4 was armed with the 2cm Flak 30 and, until 1941, came under Luftwaffe control whether it was used by the air force or army. By September 1939 the army had 195 of them. A further 1,000 were ordered and, while the chassis weren't a problem, the Adlerwerke superstructure was and there were delays. The 2cm ammunition was carried in twenty-round magazines and there were ammunition boxes on the sides (four each) and rear (two) of the vehicle. Total ammunition carried on the vehicle was 240 rounds, but a further 640 were carried – amongst other equipment – in an SdAh 95 ammunition trailer. Ramps were supplied to allow the cannon to be mounted and demounted as required. This vehicle, seen in 1940, shows Luftwaffe crew: note the flashes of *hochrot Kragenspiegel* (bright-red collar tabs) on the two men in *Fliegermütze* (flight caps). Both these men and the *Obergefreiter* at right using the rangefinder have trade badges on their left forearms: those on the vehicle probably that of AA personnel; the *Obergefreiter* probably the *Entfernungsmesser* (rangefinder) trade badge.

Inset: The German infantry squad's main weapon – the MG 34 – was a versatile multipurpose machine gun that could be carried by one man. Here, it's seen mounted on a Lafette 34 tripod in its anti-aircraft role. Note the *Trommelmagazine* – drum magazines – one on the weapon, the other providing weighted stability, the AA sight and the less than effective camouflage.

The 10.5cm leFH 18 was the standard divisional light field howitzer of the German Army. Designed with wooden wheels, the pressed-steel, rubber-tyred version was produced from 1936, although wooden wheels could be seen later in the war. The split trail gave it stability and a degree of traverse, but it made the gun too heavy for the predominantly horse-drawn battalions, especially in adverse conditions. As the photo shows (main image), even a small incline could force the crew to get off and push. The SdKfz 6 or 11 artillery tractor could tow it and its crew at up to 40kph (25mph) on a decent surface, but there were never enough to go round. Note the black-and-white aiming post stowed on the trail. This was used for indirect fire, normally planted in front of the gun at an agreed distance and used with a compass to indicate the direction of fire.

The 15cm sFH 18 heavy howitzer had a crew of seven. The 149mm x 260R shell weighed 43.52kg (95.9lb) and was propelled by a separate charge in a brass case. Both can be seen to the rear of the gun. It had limited elevation, and its maximum range of 13,300m (14,570yd) meant it was always struggling to outdistance the competition. This led to various attempts to increase the range, including a rocket-assisted shell. To the right of the gun is the limber, which allowed the sFH 18 to be towed, the barrel disconnecting from the recovery pneumatic cylinder when being transported. Without suspension, it couldn't be towed quickly when paired with an SdKfz 7, 8-ton halftrack – but it was faster than being pulled by horses. Probably the best version was its use in the Hummel SP gun. Note the two rifles mounted at the rear of the limber, the assorted baggage and the armrest for the bench-seat.

-lorses played a huge role in the artillery arm as there was so little mechanisation of the units outside the Panzer divisions. Apart from the baggage trains and supply columns, ammunition was the key element in ensuring the guns could do their job. Most of the artillery resupply from Armee or division was by horse-drawn cart. Note the man's gas cape holder and the well-made tackle.

The French had some horsed cavalry units but theirs was – unlike the Germans' – a mainly mechanised army. French cavalry saw action as mounted infantry, but also fought a number of mounted rearguard actions. However, most of their 'cavalry' units focused on reconnaissance using motorbikes or armoured cars.

An early production SdKfz 263 (8-wheel) heavy armoured radio car, with its characteristic extended superstructure to accommodate its five crew and Fu 11 SE 100 radio suite with a bedstead-frame aerial. It had front and rear drivers, steering and drive for all eight wheels and a six-forward and six-reverse gearbox. The other members of the crew were commander, machine-gunner and radio operator. Note the 15-litre triangular racked fuel can on the front mudguard and, just above it, the military symbol for signals. Command and control of the Panzer units by radio was the biggest single difference between the German tanks and those of its enemies and one of the main reasons that German tank tactics were so much more advanced.

Above: The PzKpfw III was slow into mass-production and there were only around 350 available to the Panzer divisions in 1940. On top of this, its main gun – the 3.7cm KwK 36 – wasn't effective (a report by the Allgemeines Heeresamt (General Army Office) called it 'unsuitable as an armour-piercing weapon') and its armour was easily penetrated by the French 47mm anti-tank gun. Nevertheless, the PzKpfw III had other things going for it: a three-man turret that meant the commander could command, a radio and, of course, excellent crew training. It's fair to say that it ran rings around the Allied armour. Note the profusion of extra kit on the sides and rear decking of this vehicle – named *Gneisenau* – as it makes its way through Verdun. Below the box at the right is the *Nebelkerzenabwurfvorrichtung* rear smoke dispenser.

Below: In spite of their success, the German Panzer divisions went to war with vehicles that weren't any preferable to their opponents'; indeed, many French and British vehicles were significantly better on paper. The PzKpfw II – this one an Ausf C – was armed with a KwK 30 20mm quick-firing gun which had good armour-piercing capabilities at short range, and an MG 34. It had the drawback of only having a three-man crew, which meant that the tank commander had to aim and fire the main gun as well as command the tank. The PzKpfw II chassis went on to be used for the Marder II (two versions armed either with a Soviet 7.62mm or – the better version – with a German 7.5cm Pak 40) and the Wespe SP gun.

Most units deployed towards the front by train. Germany's rail system hadn't received a great deal of attention in the 1930s – Hitler liked motorways and mechanisation – so it came as a shock when they realised midwar that they needed a lot of locomotives quickly. The resulting *Kriegsloks*, wartime locomotives, made a big difference. Nevertheless, Blitzkrieg couldn't have happened in Poland, the west or *Barbarossa* if large numbers of troops and their weapons hadn't been positioned at their start points by rail.

One thing that DRB – *Deutsche Reichsbahn* (German Railways) – got right in the 1920s and 1930s was the creation of the *Austauschbau* (interchangeable component manufacture) wagons that could be made by many manufacturers to standard tolerances. The wagon nearest the camera is a Stuttgart-type stake wagon (note name on right-hand stake) onto which the troops are loading a Maschinengewehrwagen 36. It came in two sections, the first or *Vorderwagen* (front) was for the driver; the second, or *Hinterwagen* (back), housed the *Zwillingsockel* 36 double AA mount in an *Infanteriekarren* If 5 trailer.

The Treaty of Versailles forbade Germany from building a powered air force; it didn't forbid unpowered flight. In 1922 Hermann Goering said, 'Our whole future is in the air. And it is by air power that we are going to recapture the German empire. To accomplish this we will do three things. First we will teach gliding as a sport to all our young men. Then we will build up commercial aviation. Finally, we will create the skeleton of a military air force. When the time comes, we will put all three together – and the German empire will be reborn.' In March 1933 the Deutsche Luftsportverband (DLV – German Air Sports Organisation) was established to coordinate non-military flight training and promote civilians to become air-minded in all aspects of flying.

The photos shows a primary glider, the Schneider DFS SG 38 *Schulgleiter*, a high-wing, cable-braced single-seater for basic flight training, on its two-wheeled cart. Initially, the pilot was exposed to the elements and had no instruments; later, an enclosed cockpit was created. A bungee cord propelled the craft into the air from the top of a hill, in this case (below) on Wasserkuppe, a mountain in the state of Hesse that became the home of the world's first gliding school. The DLV was succeeded by the NSFK, the *Nationalsozialistisches* Fliegerkorps (National Socialists Flyers Corps), members of which are seen (inset) training at Lübeck. Note the leather flight helmets worn by the watchers. The man in the glider wears an aluminium crash helmet.

The Messerschmitt Bf 109 entered combat in the Spanish Civil War. Its top speed of 620kph (385mph) and its firepower – one 30mm MK 108 or 20mm MG 151 cannon firing through the propeller shaft and two 13mm MG 131 in troughs on the engine cowling – made it a formidable opponent. Each Luftwaffe fighter *Geschwader* (wing) at that stage of the war had three *Gruppen* (groups), each of three *Staffeln* (flights). The black, wavy line identifies these Bf 109E-1s as being from 8. Staffel of III. Gruppe.

World War I fighter ace Ernst Udet was so impressed by the dive-bombing prowess of the USN Curtis Hawk II that he convinced the newly formed Reichsluftfahrtministerium (RLM – German Air Ministry) to fund the purchase of two to evaluate. He later used one of these Hawks to perform aerobatic shows during the 1936 Olympics. The RLM put out an industry request for a dive-bomber and the first was the Henschel Hs 123 biplane, which entered Luftwaffe service in 1936. By that time the Ju 87 was under development, its nickname coming from the generic *Sturzkampfflugzeug*, 'diving fighter', shortened to Stuka. It proved to be one of the icons of Blitzkrieg. This is II./StKG 1 commanded by Major Johann Zemsky, who was shot down over Stalingrad on 28 August 1942 and posthumously awarded the *Eichenlaub* (oak leaves) to accompany the *Ritterkreuz* (Knight's Cross) awarded on 4 February of the same year.

The Ju 87R (for *Reichweite* – range) was a Ju 87B that could carry two under-wing 300-litre (80 US gal) drop tanks. Carrying a single 150kg (550lb) bomb, it was intended for anti-shipping operations and during *Unternehmen Weserübung* Ju 87Rs sank a number of British ships, including destroyers HMS *Afridi* and the French *Bison*, as well as the sloop HMS *Bittern*. Another sloop, HMS *Black Swan*, was also hit but the bomb passed through it before exploding.

The workhorse of the Luftwaffe's transport commands, the Junkers Ju 52/3m was a trimotor with a fixed undercarriage and corrugated duralumin skin. With a crew of three or four and three MG 15s for defence, it could transport either eighteen fully armed troops, 3,000kg (6,600lb) of supplies or twelve stretchers. It was also the main mode of transport for the *Fallschirmjäger* (it could carry twelve), who exited from side doors, their supply containers dropped from the bomb-bay. It could also tow two DFS 230 gliders. As an example of its importance to Blitzkrieg one need look no further than the Norwegian campaign when 3,018 Ju 52/3m sorties were mounted, delivering 29,280 men, 2.4 tonnes (2,376 tons) of supplies and 1.2 million litres (259,300 Imperial gallons) of fuel. This photograph shows a later version of the Ju 52/3m on its way to Tunisia over the Mediterranean. Rudder tactical codes (here H2C) were brought in after 1941, the first letter being the unit group identifier (Geschwaderkennung), the number indicating Staffel, and the third the aircraft's individual Staffel identity.

KG 51 – the *Edelweiss* wing – was formed in May–December 1939 and saw action in the attacks on France and the Balkans, and in the Battle of Britain. Its main aircraft was the Junkers Ju 88, the workhorse of the Luftwaffe's bombing units of the early war period. Over 15,000 were built (second only to the B-24 Liberator in number) and it went on to be active throughout the war in various forms, including nightfighter, torpedo bomber and as a dive-bomber (although its airframe wasn't up to it).

The Luftwaffe had various ground elements during the war, none more famous than its paratroop arm. The *Fallschirmjäger* came into their own in 1940. Delivered by glider and by air, they carried out the first significant wartime use of vertical envelopment. The Germans had been impressed by the Red Army's parachutists and had set up a jump school at Stendal-Borstal in early 1936. All trainee *Fallschirmjäger* were Luftwaffe volunteers and had to meet high physical fitness requirements and to have passed rigorous courses at Stendal. There was also a school at Chemnitz, a Luftwaffe signals training establishment, where this *Fallschirmjäger* group is on a training exercise. They wear the special *Fallschirmjägerhelm* (paratroop helmet) and second-pattern smocks made from olive-green water-repellent gaberdine. It had step-in mid-thigh legs with a fly-front concealing a heavy-duty brass zip. On the outer seam of each leg was a press-stud which allowed each to be gathered in and secured. Their gloves are elasticated fine grey leather. Some men have press-stud opening map cases and Kar98k rifles, and one wears the early Luftwaffe blue paratroop bandolier with between twelve and twenty ammunition pockets. They had press-stud closings, each holding one five-round stripper-clip.

Above: The heavy cruiser *Admiral Hipper* was the lead ship of a class of five, the others being *Blücher* (sunk in Oslofjord during *Unternehmen Weserübung*), *Prinz Eugen* (saw action with *Bismarck* in 1941 and ended the war in the Baltic supporting the retreating German Army), *Seydlitz* (launched, unfinished, renamed *Weser* and conversion to a carrier started in 1942 but was also unfinished) and *Lützow*. Launched but unfinished in 1939, *Lützow* was sold to the Soviet Union and renamed *Petropavlovsk*. The class was armed with eight 20.3cm (8-inch) and twelve 10.5cm (4-inch) guns, twelve 3.7cm and eight 2cm AA guns, and twelve 53.3cm (21-inch) torpedo tubes. *Admiral Hipper* was the most successful of the class, taking part in *Unternehmen Weserübung*, during which it sank the British destroyer *Glowworm*, a one-sided action after *Glowworm* had lost its bow ramming the larger cruiser. Kapitän zur See Hellmuth Heye took the unusual step of recommending that *Glowworm*'s captain, Gerard B. Roope, be awarded a posthumous Victoria Cross; he was.

Opposite: A colourful array of flags fly from a Kriegsmarine ship, possibly *Admiral Hipper*, in harbour on the French Atlantic coast, 1941. The International Code of Signals (ICS) was ratified in 1855 and is still in use today. Each flag represents a letter or a number and a meaning. 'P', for example, is represented by a dark-blue rectangle with a white one inside and is known as Blue Peter. Its meaning: 'crew return to ship, about to sail'. The French Atlantic coastal harbours proved extremely important for the Kriegsmarine as they meant the Germans didn't have to run the gauntlet of the Allied-controlled seas between Britain and Greenland. U-boat pens were set up at Bordeaux, Brest, La Pallice/La Rochelle, Lorient and St Nazaire.

Above: Grossadmiral (Grand Admiral) Erich Raeder gives a speech in Dresden in 1939. He was shocked when Britain and France declared war on Germany as he knew that the Kriegsmarine (Wartime Navy, as it was named after 1935) was not ready and would not be so for at least five years. He wrote in the Seekriegsleitung (Maritime Warfare Command) war diary on 3 September 1939: 'As far as the Kriegsmarine is concerned, it is obvious that it is not remotely ready for the titanic struggle against England.' After the failure of the surface fleet in the Battle of the Barents Sea, Raeder offered his resignation to Hitler. It was accepted and Raeder's hated rival, Admiral Karl Dönitz, the commander of the Kriegsmarine U-boat arm, took over.

Below: The Kriegsmarine was resurrected from what had been allowed at Versailles by a major shipbuilding campaign. Hitler negotiated the 18 June 1935 Anglo-German naval agreement that paved the way for growth, but time, lack of finances to pay for raw materials and competition with the army and Luftwaffe meant that the German surface fleet was definitely the poor relation of the Wehrmacht in 1939. Nevertheless, it played a significant part in Blitzkrieg, particularly *Unternehmen Weserübung*, when it ferried troops to Norway and kept the Royal Navy from having a decisive impact on the campaign. This is Type 1934A destroyer Z12 *Erich Giese* seen in 1939 still carrying its peacetime number of 82. The first digit, 8, relates to flotilla or division and the second digit is the number of the vessel within the flotilla. *Erich Giese* took part in the Narvik landings, transporting elements of Gebirgsjäger-Regiment 139 and fought in both sea battles off the port. It was sunk on 13 April.

Kiel was a major naval base and a shipbuilding centre – 131 U-boats were built at the Germaniawerft – and heavily bombed throughout the war. *U-51* was one of the U-boats built at Kiel, launching on 11 June 1938. A Type VIIB, it had a short but dramatic career: four patrols and one wolfpack, six ships sunk, and then met its fate in the Bay of Biscay on 20 August 1940, sunk on the surface by torpedoes fired by HMS *Cachalot*.

Inset: For the first time in naval warfare, aircraft were a significant menace and anti-aircraft guns played an increasingly important role. These men are from an onboard Flak unit: the frame and netting behind them is for catching the shell cases ejected from the 2cm Flak 30/38. The four wear the six-button Kriegsmarine heavy woollen guard coat with leather shoulder trim, high collar and rolled removable hood. The coats varied in manufacture and appearance – some had slant pockets and flapped ones underneath; others lacked the flapped pockets but had turned-up sleeves. They also came in and weathered to a variety of colours as seen in the photo. Their caps are the standard Kriegsmarine navy blue *Bordmütze* (sidecap) with Bevo weave insignia above the *Reichskokarde* (the national cockade). The second man on the left has a pair of black textured binoculars around his neck. The objective barrels have long prism housings which suggests that they might be a Carl Zeiss model. On the right is the Flak gun communication officer with a boom microphone on his headset and a control unit on his chest.

parade in Warsaw. So, the vicious German occupation of Poland began. Six million Poles died in World War II, half of them Jews as part of the Holocaust, the others civilians and soldiers murdered either during the German or brief Soviet occupation.

As soon as they could, the bulk of the German Army raced back to the west and Hitler began to push for an early campaign against France.

The Polski Urząd Pocztowo-Telegraficzny – the Polish Post and Telegraph Office – in Danzig, was attacked on 1 September 1939. The Free City of Danzig had a predominantly German population but after Versailles was under League of Nations governance. The Polish minority had certain privileges, which included running the Post Office. The push to remove the 'Polish Corridor' and make Danzig German again was one of the excuses that Hitler used to justify the attack on Poland. As tensions rose, the Post Office staff were supplied with weapons and personnel intended to defend it long enough for the Polish Army to reach them. They never did. After a heroic defence by the defenders, part of the outer wall was blown out and the German assault engineers used flamethrowers, forcing the survivors to surrender. Most were executed but two survived to tell their story. The 1979 film adaptation of the Günter Grass novel, *The Tin Drum* (*Die Blechtrommel*) has graphic scenes of this event.

The Wehrmacht invaded Poland from west, north and south, each attack mixing armour with infantry mainly dependent on horses for haulage of heavy weapons and equipment. Infantry divisions advanced along several routes, with the motorised elements organised into one or more motor echelons. These followed the march column by bounds, or moved in a column along a separate road. The problems came when they were forced to move down a single road, thus extending the length of the column and the time it took to assemble for an attack. *Fliegermarschtiefe* – air marshalling depth – saw the columns move into a more open order to reduce the impact of air attack. This photographs shows a motorised column waiting to move. Seen in the photo amongst others, from the left: a Horch 4x4 Kfz 16 with canvas hood, a Mercedes-Benz WK 230 with hood down and a couple of Opel Blitz 3.6-36s, with the lead one (with the marking M) towing a 2cm Flak 38 with gun-shield. Red flags marked the front and rear of a section within a larger convoy.

The main component of the Panzer units attacking Poland in 1939 was the PzKpfw I: 40 per cent of the attacking force. Most of these were Ausf As, as here. Note the Panzer grey paint scheme (albeit covered in the dust of battle), plain white cross on the turret – replaced by the white-outline, black-core version in early 1940 – the unit identity number 535 on the plate on the side and the removed track. The tank is probably suffering from mine damage but doesn't appear to have burnt out so is probably recoverable. The standing man shows just how small the PzKpfw I was.

The Treaty of Versailles forbade Germany from producing armoured vehicles, but allowed the German police to have a small number. Needless to say, this loophole was exploited, and the result was a range of reconnaissance vehicles such as this *schwerer Panzerspähwagen* (heavy reconnaissance vehicle). Bussing-NAG, a truck and bus manufacturer, was given a contract to develop this and in 1937 the SdKfz 231 (8 Rad = 8 wheels) was introduced. It had front and rear drivers and its six gears were as fast in reverse as going forward – around 80kph (50mph). This is the SdKfz 232 version with its frame antenna and medium-range radio sets. It's passing a large yellow warning sign that advises vehicle drivers of a destroyed bridge ahead but that *'Für geländegängige Wagen umgehungen vorhanden'* – for off-road vehicles there are workarounds.

Ju 52s over Poland. The Luftwaffe's main transport aircraft, Ju 52s flew over 2,500 sorties during the invasion, the most important roles being transporting troops to their staging points and, of course, movement of supplies. Some of this involved dropping supply canisters to advanced troops. These went through bomb-bay doors.

ACHTUNG!
ZERSTÖRT!
FÜR geländegängige WAGEN
UMGEHUNGEN vorhanden!

II./Artillerie-Regiment 50 was formed with three motorised batteries, one (4.) of 10.5cm leFH 18s and two (5. and 6.) of 15cm sFH 18s. They served as corps troops of IV. Armeekorps in Poland, crossing the border east of Opole and advancing towards the Vistula. The sFH 18s were towed by the SdKfz 7 8-ton prime movers that could easily accommodate their crew of seven and, in an enclosed ammunition compartment at the rear, eighteen rounds with cased charges – more ammunition usually being carried by lorry. The wicker mats were used to line the ammo compartment. Note the travelling arrangement for the howitzers, utilising a two-wheel bogie with the spades that added stability when in the firing position carried on the gun legs.

Inset: A Horch 830 R and Mercedes L Series pause in the attack. Note shrouded headlamps on both vehicles, staff car pennant holder frame at the front, map case and gloves hanging from the dashboard, belt-order map case and bayonet with its *Troddel* (bayonet knot) and pistol holster hanging by the rear wheel, tow hook, toolbox and various containers and a folded heavy-duty canvas hood. A hint of bright red (*hochrot*) on the shoulderboard on the man at the back of the car suggests an artillery group checking a location.

Above: After a continuous seven-day aerial bombardment the siege of Warsaw started on 8 September 1939 and lasted until the 28th, when the garrison of approximately 140,000 surrendered. The Germans entered the city on 1 October 1939 and stayed for over four years until 17 January 1944, when it was liberated by the Red Army. Warsaw was slated for demolition and rebuilding as a great new German city. It hadn't happened by 1944 but in an act of wanton destruction the Germans razed 80–90 per cent of the city in the last months of their occupation, targeting anything cultural or historical – libraries, churches, castles and museums – looting as they destroyed. The city had to be rebuilt after the war. The 1939 photo shows a 3-ton Opel Blitz truck carrying troops, with more in the two-axle mid-rise box trailer.

Opposite: A German soldier plays for the camera at a shooting range in Poland. He's wearing a reversed Polish four-cornered cap – a *czapka*, probably cavalry – a local shirt and grubby vest, regulation trousers and braces, rag-covered footwear, his Kar98k rifle and bayonet and an ammunition pouch.

Krasiński Square, Warsaw, in 1940, a strange mixture of calm and destruction watched over by the statue of Jan Kiliński, a nineteenth-century rebel. It was designed by Stanisław Jackowski and unveiled on 19 April 1936. The Germans moved the statue in 1942 and were going to destroy it, but luckily the efforts of the director of the National Museum in Warsaw, Stanisław Lorentz, saw it stored instead. It's one of the few pieces of prewar Warsaw to have survived the German devastation and currently stands on Podwale Street in the Old Town.

gland"

This photo was taken in Poland around the time of the attack on the Soviet Union and reinforces the fact that the German Army was primarily horse-drawn. The army had a range of Heeresfeldwagen (army field wagons), the most numerous being the Hf 1 drawn by two horses and the Hf 2 by four. Between 1940 and 1943 the German Army requisitioned 1.2 million horses from Germany and the occupied territories – particularly Poland. They needed to because during the same period they lost 1.5 million horses.

Luftwaffe Hauptmann Rolf mit Russdorfer of Flakgruppe *Memmingen*, in Bavaria in winter 1939–40. The airfield there was the home of Kampfgeschwader 51, a bomber unit mainly equipped with Do 17Ms and He 111Hs. He has *hochrot* (bright-red) *Waffenfarbe* on shoulderboards and collar *Kragenspiegel*, whose silver-embroidered wreath and single gull identify his rank. On his head he wears an officer's *Fliegermütze* with officer's silver-aluminium piping. Note the heavy roll-neck jumper under his coat. During the *Sitzkrieg*, Flak units had little to do as most Allied aircraft were either on leaflet-dropping or reconnaissance missions. Flak had been sited near airfields as the OKW expected them to be bombed and much prewar doctrine concerned the creation of a Flak barrier between the Western Allies and Germany. In 1940, particularly after the armistice with France, that changed to protecting industrial centres.

Dragon's teeth – in German *Drachenzähne* or *Höckerlinie* – were pyramidal reinforced-concrete structures of many shapes and sizes with deep foundations, laid between fortifications, bunkers and pillboxes, often mined and with barbed wire and other obstacles to impede infantry. Most often seen in multiple lines and behind substantial minefields, they were designed specifically to disrupt tank advances, forcing them into killing zones and interlocking anti-tank fire. Stretching from Kleve on the border with the Netherlands south to Switzerland, the Germans spent much time and money on these defences in the 1930s, but they weren't really tested until 1944.

MG 34 practice for Artillerie-Regiment 205 on the Westwall in 1939. Part of 205. Infanterie-Division under Generalleutnant Ernst Richter, the unit trained and worked from its creation in January 1940 till its entry into battle in France in June of that year. Until then, 205. Infanterie-Division manned the Westwall. With 18,000 bunkers, tunnels and tank traps it was, on paper, a formidable barrier. In 1944 it proved less so, having been stripped of almost everything that would be of more use in the east. During the *Sitzkrieg* the Westwall was a peaceful training area and suitable for observation across the border. The only minor skirmish was in the Saar

area when the French Army made a half-hearted incursion following the German attack on Poland. The MG 34 was a recoil-operated, air-cooled, machine gun firing a 7.92mm rifle cartridge via an open bolt feed o non-disintegrating reusable metal link belts. A select-fire weapon, semi-automatic or fully automatic, it was capable of firing 900 rounds per minute – although in practice it was best in short, controlled bursts that didn't melt the barrel. The steel ammunition container, as seen in the photo, held 250 linked rounds. Note how the firers use their left hand to pull the butt into their shoulders. The

Artillerie-Regiment 205 guard carrying a Kar98k rifle in a dugout in Westwall. He wears an M36 double-breasted field-grey wool greatcoat with dark-green collar, large slanted outer pockets and adjustable-length cuffs. His Kar98k woodwork looks to be walnut rather than laminate – as produced up until 1940. Laminate was trialled in the late 1930s and proved more robust and durable than hardwood. Note the flimsy effort for concealment and the brand-new duckboard.

The German Flossacke 34 inflatable was designed for river crossings and assault, carrying eight men or more. If under fire, it was rowed from a crouching position with one leg outside and the other in. When under fire the inflatable offered a perilous crossing with no protection to the occupants. The rope on the outside was for holding onto if you fell overboard. The boats would be carried to the riverbank already inflated, then occupied and pushed off. The soldiers shown are familiarising themselves with the boat while under supervision from the instructors at the stern. The assault pioneer boat at the back has an outboard motor and is handled by a *Steuermann* (helmsman) whose arm badge would identify his specialist qualification.

Time was of the essence for every Flak gun crew as a few minutes, even seconds, made the difference between a hit or a miss. Optical instruments were the eyes of the Flak troops: gun sights, spotting binoculars and, as seen in the photo, rangefinders. Putting it simply, in order to shoot down a high-flying aircraft, gunners had to know three parameters: altitude, direction and speed. In theory, it was a simple matter of three-dimensional geometry. The observer looks through the double eyepiece and adjusts a fixed mark on the target and by rotating the prisms at either end until the mark appears to overlap in the observer's view. By measuring the angles of the prisms to the target and the known width of the baseline (4m, 12ft) the range is calculated. The highly trained crew – range-taker (observer), layer for line, layer for elevation and a reader – would provide that information from their stereoscopic rangefinder, in this case a Zeiss EmR 4m, the standard instrument for the heavy guns such as the 8.8cm Flak.

Of course, that was only the range! Concurrently, the *Kommandogerät* (stereoscopic director or predictor) – a fire-control computer with a crew of five – was number-crunching all data into the predictor giving instant data of speed, direction and height to the gunners. The white overalls are fatigues, standard in the German Army.

German soldiers receiving training in the use of the GM 30 gas masks: how to put them on and take them off – probably timed as well – how to maintain the masks, the sturdy waterproof container and its contents: cleaning cloth, a small container with anti-fog lenses and spare lenses in the lid. There's decontamination training going on along the treeline. All German personnel and civilians received a gas mask and training. When the soldiers got back to the barracks, the masks were removed from the container because they would become deformed if left there. They would be stretched over an aluminium and sprung wire frame – *Maskenspanner* (gas mask storage spring) – to keep the correct form.

Luftwaffe SdKfz 7 halftracks tow 8.8cm Flak guns past a Goldene Sieben shop in Halle. A gas guzzler at 1.3km/l (3mpg) on the flat and 0.64km/l (1.5mpg) cross-country, the SdKfz 7 carried twelve men, including the driver, sitting on three bench seats. Ammunition, supplies and equipment were in storage compartments at the rear, and it had a 3.8-tonne (7,600lb) winch for unhitching itself or other vehicles. The 8.8cm (nicknamed 'Acht-acht') Krupp-designed and -built Flak 18 being towed is on a six-wheeled (two at the front; four at the back) SdAh 201 trailer. Note the WL numberplate (Wehrmacht Luftwaffe) and crew with the bright-red *Waffenfarbe* collar tabs of Luftwaffe Flak units.

Goldene Sieb

Above: The 5cm leGrW 36 (*leichter Granatwerfer* – light mortar) smooth-bore, muzzle-loaded, trigger-fired (percussion) mortar had a two- or three-man crew although one man could operate the weapon in an emergency. It weighed 14kg (30.8lb) and only fired high-explosive rounds and smoke from 60m (65yd) to 500m (547yd). With experience an operator could fire by the angle of the barrel rather than by the earlier complicated telescopic sight. It was withdrawn from the front line in 1942. The mortar broke into two components – baseplate and firing tube – and a special metal frame allowed it to be carried on soldiers' backs. This is visible between the *Zeltbahn* quarter roll and his waist.

Opposite above: An artillery radio team train on their TornFu f (Torn = *Tornister* = backpack; Fu = *Funkgerät* = radio). Introduced in 1936, this is a combined transmitter (on the left, S = *Sender*) and receiver (on right, E = *Empfänger*) which were transported in a case weighing 20kg (44lb) – about the same as the similarly sized battery accessory case which carried the antenna parts, batteries, headphones, key, microphone and other accessories. Each man would carry a case. The man with the headphones is setting the control switchbox that has three settings: *Ortsbesprechung* (local operation), *Fernbesprechung* (remote operation) and *Zwischen Verkehr* (telephone operation). Note the Bakelite microphone next to his left hand. The red button has *Drücken* (press) on it for use and one of two admonitions: *'Feind hört mit!'* or *'Achtung! Feind hört mit!'* (Beware! The enemy is listening).

Opposite below: Schützen-Regiment 103 was based at Lobau Barracks in Saxony in 1939. *Schützen* is an old military term for shooters or marksmen. These recruits, wearing their parade white work uniform, will become motorised infantry as part of 14. Panzer-Division. Note the scruffy soldier saluting an NCO in the earlier and correct manner.

Recruits from Schützen-Regiment 103 train on the 7.5cm leIG 18 – an infantry close-support howitzer. It had a shotgun breech mechanism. All the loader had to do was lift the rear of the barrel clear of the block, load and lower it until the breech was closed. Its maximum range was 3,375m (3,690yd) and it fired high-explosive and a hollow-charge shell for tanks and enemy personnel. The two men behind the shield have towing straps over their shoulders; one hook is visible and with two others the howitzer can be towed. The spoked wheels were later replaced by pneumatic tyres.

A turretless training PzKpfw II on a Wehrmacht training ground in 1940. During the early years, Panzer training was extensive but this was reduced to twelve weeks due to the increased demand for tank crews. At that point they put greater emphasis on battle drills, digging in and the techniques of camouflage – all skills that needed reinforcing after heavy battle losses.

	to 1938	1938	1939	1940
SdKfz 263 (8-wheel) radio ac	–	64	20	–
Austrian Steyr ADGZ (8-wheel)	–	[14]	–	–
French Panzerspähwagen P204(f) (Panhard 187) (4-wheel) inc some radio ac	–	–	–	[233]
Total	*707*	*211 [14]*	*324*	*189 [233]*
Prime movers				
SdKfz 2 klKettenkrad[1]	–	–	70	500
SdKfz 6 mZgkw 5t[2]	?	?	1,751	617
SdKfz 7 mZgkw 8t[3]	985	226	1,609	996
SdKfz 8 sZgkw 12t[4]	?	?	780	516
SdKfz 9 sZgkw 18t[5]	2	38	?	?
SdKfz 10 leZgkw 1t[6]	?	11,300	?	?
SdKfz 11 leZgkw 3t[7]	?	15,000	?	?

Notes

Exact figures of prime movers difficult to establish

1 70 produced 1939–40 testing series; 500 produced 1940–41.

2 1,751 produced 1934–39; 617 produced 1939–41.

3 985 produced 1934–37; 12,187 produced 1939–45.

4 780 produced 1934–39.

5 2,591 produced 1939–44.

6 11,300 produced to 1939; 14,698 1939–45; c.7,500 converted to SdKfz 250.

7 9,028 produced 1939–45; from 1939 c.16,000 converted to SdKfz 251.

Above: A rare vehicle (only seventy-five were produced), this is a PzKpfw II Ausf a – not to be confused with the Ausf A, the final standardised version with five larger wheels, no girder and four return rollers. The small wheels of the Ausf a were sprung in three pairs under three bogies supported by a reinforcement beam and three return rollers. The rear end of each fender could be raised to prevent the buildup of debris and mud. The hand-turned turret, without a cupola at this stage of design, employed the new KwK 30 L/55 20mm quick-firing gun which had good armour-piercing capabilities and was fed from a ten-round box magazine, and an MG 34. The crew of three was commander/gunner, driver and radio operator/loader under the turret with a *Funkgerät* – FuG 5 USW receiver and transmitter. Note the engine compartment at the rear is covered with clutter, a regular cause of overheating.

Opposite: The light PzKpfw I Ausf A was the first German tank produced in large numbers and saw action well into 1941 in its command version. Underpowered and prone to overheating, it was, nevertheless, an important element of Blitzkrieg. It's difficult today to understand how something this size could revolutionise warfare, but it was a major component of the Panzer units in the early war. Armed with two 7.92mm MG 13 in a hand-turned turret, its main role was to support infantry, acting as a mobile MG nest. To get out of trouble it could lay smoke from the NKAV smoke dispenser seen above the towing cable at the rear of the tank.

The PzKpfw II Ausf B had a reworked flat hull, five independently sprung larger road wheels and four return rollers. A small commander's cupola was added as well as additional armour bolted to the front of the welded hull. Just in front of the cupola is a small dummy periscope that was retained on some models but not others. The main armament was a 2cm KwK 30 L/55 cannon with a ten-round box magazine. It fired two types of ammunition – armour-piercing or high explosive. The motorbike is a BMW R Series.

The PzKpfw II Ausf B had a reworked hull front with additional 20mm (c.0.8 inch) armour bolted to the now angular glacis and around the driver's vision port, which had 50mm (c.2 inches) of bulletproof glass held by two external bolts. An eight-periscope cupola was added, increasing the commander's vision, and the dummy periscope in front of the cupola was removed. Ammunition for the KwK 30 had improved: the Panzergranate (PzGr) 39 armour-piercing round could penetrate 23mm (c.0.9 inch) at 100m (109yd) and the PzGr 40 APCR (armour-piercing composite rigid) 40mm (c.1.6 inch) at 100m; it also fired a Sprenggranate (Sprgr) 39 high-explosive round. Secondary armament was an MG 34. The earlier six wheels were replaced by five larger ones and four return rollers rather than three. The crew was the same as earlier versions. Note the spare road-wheel and track added to the glacis plate for extra protection. The following Panzers are earlier PzKpfw IIs with the rounded hull front.

Inset: The medium PzKpfw 38(t) was an important stopgap for the German Army as production of the home-grown PzKpfw III and IV took longer than hoped. Produced by the Czech company KD, it was introduced into German service as the LTM 38 and performed well in Poland. On 16 January 1940 it was redesignated PzKpfw 38(t) – where t = *tschechisch* = Czech. Over 1,400 were built for the Germans, with one significant change: the single-man turret was altered to allow a loader's position by reducing ammunition capacity by eighteen rounds. Reliable and extremely mobile, after it was replaced by more modern vehicles the chassis was used for a variety of different vehicles, including the Marder III *Panzerjäger* and, later, the Jagdpanzer 38(t). This photograph shows a 38(t) in Poland. Note the commander is wearing the black Panzer uniform and unpopular M34 beret (*Schutzmütze*) and the vehicle has the solid white *Balkenkreuz*.

Disabled PzKpfw 38(t) – the flag is the yellow and black *Ausfallflagge* (breakdown flag) – an unusual sight: the PzKpfw 38(t)s were well known for their reliability. During the campaign, the Germans lost a third of their tanks – over 800, including fifty-four PzKpfw 38(t)s – and there were some 1,700 that required heavy repair. However, they had done their job and spearheaded the Blitzkrieg.

The PzKpfw III Ausf E armed with a 3.7cm KwK L/46.5 gun and two 7.92mm MG 34s, one in the turret and the other in the Kugelblende 30, hull machine-gun housing (although this tank's MG isn't deployed). The oak leaf symbol on the hull indicates that this tank is from 1. Panzer-Division which, as part of Guderian's XIX. Panzerkorps, had broken through at Sedan – a critical point in the French defences – and then motored

across the country to Abbeville and on to Boulogne. During *Fall Rot*, the division attacked in the east, crossed the Marne at Châlons, advanced to Besançon and ended the campaign in Belfort. The number 341 on the turret could well indicate 3 = company, 4 = platoon and 1 = position in platoon.

The PzKpfw IV Ausf D was armed with 7.5cm KwK 37 L/24 gun and was originally intended as an infantry support tank. However, the generous size of its turret ring made the PzKpfw IV more easily upgunned and it became the workhorse of the Panzerwaffe. Note the distinctive muffler with exhaust pipe over its top, wide hinged fenders and towing bracket with pin below. Above the muffler is the NKAV smoke dispenser.

Inset: As a footnote, the most numerous and effective German AFV of the war was the *Sturmgeschütz*. It cannot be called a primary weapon of Blitzkrieg, because its first significant campaign was *Barbarossa*. Some thirty saw action in France in 1940 and it proved extremely useful in its intended role as a mobile infantry support weapon – its tank-killing days coming later. This one was supporting 57. Infanterie-Division.

5 *Unternehmen Weserübung* – The Attack on Denmark and Norway

GROSSADMIRAL ERICH RAEDER HAD been lobbying since the war began for an invasion of Norway to facilitate U-boat operations against Allied shipping. The air bases would also help in any campaign against Britain itself. Hitler was more worried about the safe passage of iron ore-carrying vessels that used the Norwegian channel through the Skjaergaard – the myriad islands along the coast of Norway – which was susceptible to mining. The *Altmark* incident on 16 February 1940, when the Royal Navy intercepted a German ship carrying PoWs in Norwegian waters, concentrated minds and Hitler's directive for *Unternehmen Weserübung* was sent out on 1 March:

> 1. The development of the situation in Scandinavia requires that all preparations be made to occupy Denmark and Norway with parts of the Armed Forces. This measure should prevent the British from gaining a foothold in Scandinavia and the Baltic Sea, safeguard our ore resources in Sweden, and expand the bases of the Navy and the Air Force for attacks on the British Isles.
> Insofar as possible, the Kriegsmarine and the Luftwaffe have the task of protecting the operation against interference from British naval and air forces.
>
> In view of our military and political strength in relation to the Nordic states, the forces to be used for *Weserübung* should be kept as small as possible. The small numbers must be compensated for by bold action and surprise. …
>
> 2. I order General von Falkenhorst (Infantry), Commanding General of the XIX. Army Corps (Commander of Group XXI), to make preparations and command the operation against Denmark and Norway. …
>
> 3. …The most important thing is for our operations to take the Nordic countries as well as the western enemies by surprise. This should be kept in mind in making all preparations, especially in assembling the troops and the shipping space, training and loading. …
>
> 4. Occupation of Denmark (*Weserübung Süd*)
> Task of Group XXI: A surprise occupation of Jutland and Fuenen, and subsequent occupation of Zealand. …
>
> 5. Occupation of Norway (*Weserübung Nord*)
> Task of Group XXI: Surprise occupation of the important coastal points from the

sea and air. … Cover names: *Wesertag* – day of the operation (9 April); *Weserzeit* – hour of the operation (04:15).

The actions of the Wehrmacht in 1939–40 were risky. The attack on Poland with the bulk of German troops could have been derailed by an Allied attack in the west. *Fall Gelb* was predicated on the surprise attack through the Ardennes, which, if it had been spotted early enough, could have ended in stalemate. And *Unternehmen Weserübung Nord* was an operation in the face of the Royal Navy, reputedly the dominant force in the area – well, that's what both Norwegian and British military establishments felt. They also thought that the main German threat would be an attempt to break the blockade: to challenge the Royal Navy with capital ships.

The Germans fooled them, sent a number of small invasion fleets using warships rather than slower troopships, and in a spectacular tri-service operation landed troops from both air and sea. Denmark fell quickly; Norway in a campaign lasting two months. *Weserübung Nord* saw the first significant use of *Fallschirmjäger* – a battalion's worth of paratroops – along with over seven divisions of infantry, most of the Kriegsmarine, and over a thousand aircraft of Luftflotte 5.

The key to the successful operation turned out to be the Luftwaffe's Fliegerkorps X rather than the Royal Navy. Commanded by Generalleutnant Ferdinand Giesler, it delivered *Fallschirmjäger* to key locations and then resupplied them, intercepted Royal Navy assets at sea and bombed the ground forces on land. This Luftwaffe had set up special-purpose wings *Kampfgruppe zur besonderen Verwendung/Verfügung* = combat group for special deployment/duties) to cater for air transport: KGr zbV 101–107 equipped with the 'Tante Ju' – Ju 52/3m trimotor aircraft. With over 300 bombers – including two wings of He 111s, one of Ju 88s and a group of Stukas – two groups of Bf 110s and one of Bf 109s, the Luftwaffe was able to command air superiority against weak RAF and Fleet Air Arm numbers.

The Germans didn't have it all their own way: the Norwegians sank the heavy cruiser *Blücher* in Oslofjord with great loss of life, giving time for the Norwegian royal family to escape; the Royal Navy took sufficient toll on the Kriegsmarine's warships – two light cruisers and ten of its twenty destroyers – to ensure that it would not be able to support an invasion of Britain and, in the short term, the Kriegsmarine wasn't able to disrupt Operation *Dynamo*, the extraction of British and French forces from Dunkirk.

The capture of Norway proved a long-term problem for the British, providing airfields for German bombers and maritime aircraft, and safe harbour for German vessels; it secured Germany's supply of the iron ore so crucial for its steel industry, but it would also prove a drain on German resources in terms of a large occupying force and the Atlantic Wall installations built to stave off an Allied invasion.

Denmark was unable to put up much resistance to the overpowering German invasion. Casualties were light on both sides and the royal family elected to stay in the country because the crown princess was heavily pregnant and couldn't be moved. The photograph shows a III./Artillerie-Regiment 677 bus in Denmark in 1940. Buses may seem a bit out of place in war. Unarmoured, unstable off-road, they appear to epitomise

peacetime. In fact, the Germans made great use of them and buses were in the KStNs of armoured divisions as motorised field hospitals and post offices. There were three weight categories: light (up to fifteen seats); medium (sixteen to thirty); heavy (more than thirty). Many were requisitioned in large numbers from civilian operators and others were acquired in the occupied countries.

As the photograph shows, the 15cm Kanone 16 heavy field gun was transported in two sections: the barrel separated from its recoil spring. The Treaty of Versailles restricted the weapons that Germany could possess and many of the artillery pieces were given away as reparations. The Kanone 16 was no exception, although

some were retained and were used for training in the interwar period. A few were still in service in 1940 and those that had gone to Belgium were reacquired after the country surrendered. Rather than keeping their original designation, they were designated the 15cm K 429(b).

Repair or maintenance work to the rear brakes and suspension of this BMW R51 or R66. The vehicles belong to III./Artillerie-Regiment 677 (part of 11. Schützen-Brigade). At the start of April 1940, the brigade was training on the Lüneburger Heide but on 9 April it crossed the Danish border at 05.00. It was ordered to take Aalborg (an important airfield) by the next morning and also, the same day, Frederikshavn, Skagen and Hirtshals on the north coast (the latter having cable connection to Norway and England). It did so, cut the cables and set up a watch post in case the Allies tried a landing. Replaced by elements of the 160. Infanterie-Division, on 24 April the brigade began to return to Germany for rest and refit.

W.H.
M 62615

Carpool in Denmark: a *Gefreiter* with red (= artillery) *Waffenfarbe* poses by a Horch 930 V. Note the artillery pennant (black and red) on the Horch and the markings on the mudguard of the next-door Opel Olympia at right, indicating the Stab/Artillerie-Regiment 677, a unit employing leFH 18 (mot) howitzers – 'mot' meaning pulled by a tractor unit. Both vehicles have blackout covers on the headlights and radiator grille covers.

Onboard the *Isar* heading towards Oslo, an army field-kitchen (nicknamed the *Gulaschkanone* or goulash cannon in World War I, it is said because the stovepipe resembled a cannon) supported the ship's galley in feeding the passengers. It was an efficient means of providing large quantities of nourishing hot meals using the minimum of resources. Note the square red-and-white life raft, *Rettungsfloss*, with interwoven webbing floor, paddles and rope grab rails.

The U-boat depot ship *Isar* transporting troops from Hamburg to Oslo, May 1940. Built as *Puma* in 1930, the one-time fruit steamer used to carry bananas from Cameroon to Germany. The lack of vessels and need for quick remedies saw *Puma* converted in 1938 and armed with one 8.8cm, four 3.7cm and sixteen 2cm Flak guns and equipped with a submarine signalling system, wireless direction finding and an echo-sounding device.

Kristiansand was one of six primary targets attacked by the Germans on 9 May 1940. Gruppe 4 consisted of the light cruiser *Karlsruhe* and other smaller vessels carrying just over 1,000 men. The other Gruppen attacked Narvik (1), Trondheim (2), Bergen (3), Oslo (5) and Egersund (6). Gruppe 4's attack was held up by the fortifications at Kristiansand but by the end of the day the Germans were in control of the city and the island of Arendal where the undersea telegraph cable to the UK came ashore. The photograph shows follow-up troops arriving.

MALGACH

The Norwegians are people of the sea and those boats – fishing boats in particular – not captured by the Germans played a significant role in the resistance against Nazi occupation. While most of the naval ships had been sunk, the large and modern merchant fleet was a major asset to both the Norwegian government in exile and the Allies. They were also involved in clandestine operations, such as the 'Shetland Bus', a special operations group (Norwegian branch Special Operations Executive) which operated between Scotland, the Shetlands, the Orkney Isles and German-occupied Norway.

6 *Fall Gelb* – The Attack on the West

On 9 October 1939, Hitler sent out his Directive No. 6 for the Conduct of the War:

> 1. If it should become apparent in the near future that Britain and, under her influence, France also, are unwilling to end the war, I am determined to go over to the offensive without much delay.
>
> 2. A long period of waiting will result not only in losing Belgian and perhaps also Dutch neutrality, to the advantage of the Western Powers, but also in strengthening the military power of our enemies to an increasing degree; it will shake the confidence of neutrals in a final German victory, and will discourage Italy from entering the war on our side.
>
> 3. I therefore issue the following orders for the conduct of military operations:
> a. Preparations are to be made for an offensive at the northern flank of the Western Front, through the area of Luxemburg, Belgium and the Netherlands. This attack must be carried out as soon as possible and with the strongest possible forces.
> b. The purpose of this offensive is to defeat as large a portion as possible of the French field army, together with the allies fighting at her side and at the same time to conquer as large an area as possible in the Netherlands, Belgium and Northern France to serve as a base for air and sea warfare against Britain which would give promise of success, and as a wide protective belt for the vitally important Ruhr area.
> c. The date of this attack depends on the operational readiness of the tank and motorised units and on weather conditions. The forces mentioned are to be made ready for action as quickly as possible.
>
> 4. The Luftwaffe will prevent the Anglo-French air force from attacking our Army, and will, as far as is required, directly support the Army's advance. In this connection it will be important to prevent the Anglo-French air force from gaining a foothold in Belgium and the Netherlands, as well as to prevent British troops from landing there.
>
> 5. The Kriegsmarine must make every effort directly or indirectly to support the operations of the Army and Luftwaffe during the attack.

> 6. Apart from preparations for beginning the attack in the West according to plan, the Army and Luftwaffe are to be ready, at all times, and in steadily growing strength, to oppose at a moment's notice and as far as practicable any Anglo-French entry into Belgium; likewise to occupy as great a part of Holland as possible in the direction of the west coast.

On 20 November 1939 the codewords for go/no go were identified as: 'Rhein' (go through with the attack – later changed to 'Danzig') or 'Elbe' (stop the attack – later changed to 'Augsburg'). On 13 December 1939 the plan incorporated 7. Fliegerdivision ahead of the front of the 6. Armee and ahead of the 4. Armee to help take the Meuse crossings.

On 10 January 1940 Major Helmut Reinberger's aircraft made a forced landing at Mechelen-sur-Meuse in Belgium. Reinberger was commandant of the Fallschirmjägerschule (Paratroop School) Stendal and was helping with the planning for the airborne invasion of the Netherlands. He was carrying a draft of the plans for *Fall Gelb* along with detailed maps.

For the Germans, this was a catastrophe of the highest order and, two days later, Alfred JodI – Chef des Wehrmachtsführungsamtes im OKW (Chief of the Wehrmacht Command Office in the OKW) – telephoned the Chief of the Army General Staff to stop all forward troop movements. The attack, which had already been delayed on a number of occasions since being drawn up 1939, now compromised, was postponed again.

The Allied reaction to the disclosure of the German plans was muted: in Belgium, most immediately affected by *Fall Gelb*, the government ordered the penultimate step of mobilisation and called up more troops. The British and French reaction was that their own response catered adequately with the Schlieffen-style attack they had anticipated and was clearly delineated in the plans. The German response to this catastrophe changed the outcome of the war.

The *Fall Gelb* plan had been devised by General Franz Halder and the OKH in response to Hitler's decision to attack the west. Its main thrust was in the north with Heeresgruppe B (composed of 4., 6. and 18. Armeen commanded by General Fedor von Bock) being the largest component – employing thirty infantry divisions and the bulk of the mobile forces (nine armoured and four motorised divisions). Protecting his flank would be Heeresgruppe A (12. and 16. Armeen with twenty-two divisions under General Gerd von Rundstedt), with Heeresgruppe C (1. and 7. Armeen – eighteen divisions under General Wilhelm Ritter von Leeb) pinning down troops in the Maginot Line.

This plan – essentially a rerun of the opening attack in World War I – had had its critics, the most vociferous being the chief of staff of Heeresgruppe A, Generalleutnant Erich von Manstein, and his boss, Generaloberst Gerd von Rundstedt.

In *Lost Victories* Manstein outlined his point of view:

> Significantly enough, my first reaction to the plan laid down … was emotional rather than intellectual. … I found it humiliating, to say the least, that our generation could do nothing better than repeat an old recipe, even when this was the product of a man like Schlieffen. What could possibly be achieved by turning

> up a war plan our opponents had already rehearsed with us once before and against whose repetition they were bound to have taken full precautions? [Additionally, the] 1939 operation plan … contained no clear-cut intention of fighting the campaign to a victorious conclusion. Its object was, quite clearly, partial victory (defeat of the Allied forces in northern Belgium) and territorial gains (possession of the Channel coast as a basis for future operations). .... The operation planned by O.K.H. would bring partial victory, nothing more.

Manstein had been chief of the operations and plans section of the German Army General Staff, so he knew what he was talking about, and in Gerd von Rundstedt he had a professional and experienced commander. Manstein's proposal changed the emphasis of the attack to a daring armoured thrust through the Ardennes that would slice through the weakest part of the defences – weakest because the French and most other military men thought the Ardennes if not impassable then certainly difficult terrain for an army to manoeuvre through.

The proposal initially fell on deaf ears at OKH and was rejected a number of times. (It's worth pointing out that Manstein and Halder were not the best of friends.) In the end, Manstein was promoted away from the west to command XXXVIII. Armeekorps being set up in Germany. But Hitler had heard of Manstein's revisions to the plan, had doubts about the existing version and when Manstein reported to Hitler as a new corps commander, the Führer asked for his views.

Manstein gave him both barrels: first, the strategic concept of the plan was wrong. It only aimed 'to defeat as large a portion as possible of the French field army, together with the allies fighting at her side and at the same time to conquer as large an area a possible in the Netherlands, Belgium and Northern France to serve as a base for air and sea warfare against Britain'. It should, Manstein argued, 'be directed towards winning a final decision in France and destroying France's resistance'. Next, the main thrust should be by Heeresgruppe A and, to accomplish its task, it must consist of three armies. Another army must therefore be inserted on its northern flank. Finally, XIX. Armeekorps by itself would not supply enough oomph to cross the Meuse at Sedan.

Shortly after this meeting, a new final operational order was produced. In *Lost Victories* Manstein suggests Winston Churchill coined the term 'Operation *Sichelschnitt*' (sickle cut) for the attack after the war; the Germans at the time called it the fourth iteration – Aufmarschanweisung Nr 4 (Deployment Instructions No. 4) – of *Fall Gelb*. The changes to the original plan were significant, with Heeresgruppe A being beefed up to four Armeen – 2., 4., 12. and 16.– leaving Heeresgruppe B with 6. and 18. Armeen; Heeresgruppe C with 1. and 7. Also important was the fact that Heeresgruppe A had three *Panzerkorps*: XV. (Hoth) with 4. Armee and Panzergruppe von Kleist – XIX. (Guderian) and XLI. (Reinhardt) Panzerkorps – with 12. Armee.

Had the plan not changed, had the weight of the German attack been on the right flank with Heeresgruppe B, it would have run into the cream of the British and French forces head on in a prepared defence: the BEF and the French 1re Armée. Would that have made a difference? Possibly not; such was the remarkable power and speed on the German attack that it's probable that the original *Fall Gelb* may well also have been as successful although the likelihood is that it may not have been as decisive and may have led to a longer campaign.

But the plan did change and the rest is history. The final *Fall Gelb* plan was a subtle mix of feint and bluff. The feint attack was in the south where Heeresgruppe C pinned down the defenders of the Maginot Line and kept their eyes away from the Ardennes. The bluff came in the north where airborne forces and Heeresgruppe B attacked strongly into the Netherlands and towards Belgium. These attacks convinced the Allies that the main thrust was where they expected it to fall and they rushed headlong into the French Plan D, advancing into Belgium to meet the German thrust. Fixed in place, they could do little when news of the main German thrust – Heeresgruppe A spearheaded by Panzergruppe Kleist – which was held back until the Allies were committed in Belgium, reached them.

It did so, just after the Allies, particularly French 1re Armée, had beaten the Germans at Hannut and Gembloux, inflicting a major reverse and serious tank losses on German XVI. and IV. Armeekorps. However, as Jeffery A. Gunsberg put it in a perceptive article: 'Allied success at Gembloux was nullified by German victory further south.' After threading their way through the forests, steep hills and deep valleys of the Ardennes – something the heavier armour of 1944–45 couldn't do successfully – and punching across the Meuse, Heeresgruppe A's Panzers were used en masse, moving with great speed and skill, the infantry following on as quickly as they could. The Allies in the north were forced to retreat and were caught between Heeresgruppen B and A. They were annihilated.

The brilliance of the German execution of *Fall Gelb* still takes the breath away today. Between 10 and 20 May the Germans succeeded in destroying the Allied defence, forced Britain to begin to withdraw its troops from the continent and set up France for a final dismemberment. By 4 June Operation *Dynamo* and neighbouring evacuations had rescued c.221,500 British and c.170,500 French troops, but they'd left behind them thousands of other troops and all their transport and heavy weapons; many of the latter – particularly the lorries – would see German service in the coming months.

Belgium

German transports were heavily involved in *Fall Gelb* and, as well as the Eben Emael glider landings, there were attacks on routes into Belgium through Luxembourg. *Unternehmen Niwi* saw III./*Grossdeutschland* Regiment delivered by aircraft – mainly Fieseler Storch – to clear the way for the Panzer divisions attacking through the Ardennes. The 'Tante Jus' suffered significant losses during the campaign in the west, with estimates suggesting that between 275 and 350 were destroyed or heavily damaged. The factors involved included accidents, enemy action and heavy landings.

Inset: An important element in the attack on the Low Countries was the destruction of the air forces. The Luftwaffe attacked the airfields of the Aviation Militaire Belge early in the morning of 10 May and knocked out around eighty aircraft. This is one of the bombed Belgian military airstrips, showing the destruction to the hangars. The vehicles seen are German petrol trucks.

The Belgian defences in the east along the Dyle Line were anchored to the fortifications around Liège and Namur. The German attack south of Namur broke through between Dinant and Sedan and saw the French 9e Armée destroyed. This led to a withdrawal to the Escaut Line. Namur, isolated, was abandoned and the forts fell one by one: first Marchovelette on 18 May and the last, Maizeret, on the 23rd. This photo shows the bridge over the Meuse at Dinant which was blown up on 12 May. It didn't stop Rommel from forcing a crossing the next day.

VILLE DE BRUXELLES
FABRIQUE DE COUQUES
TAVERNE

Pretty much due west of Dinant on the Franco-Belgian border lies Beaumont. On 16 May, the French 37e BCC (bataillon de chars de combat – tank battalion) part of 1re DCR (division cuirassée – armoured division) was forced to withdraw along with the rest of 11e corps d'armée of 9e armée. 37e BCC – equipped with B1 bis heavy tanks – was stranded in Beaumont without fuel. There was little choice but to destroy the tanks to

stop them falling into enemy hands. The effectiveness of this procedure is exemplified by the charred remains of *Bearn II* on rue Madam. On its side, the lozenge markings, number 444 and message *'Schrott'* (scrap) indicate that *Bearn II* had been assessed by Panzerbeutestab 2 and scrapping was the best course of action (for more on the *Panzerbeutestäbe* see p. 238).

A bombed-out port bridge in Portus Ganda, the old port area of Ghent (Ganda = Ghent in Latin) at the confluence of the rivers Lys and Scheldt. Ghent was a strategic target for the Germans because it was on a vital waterway connecting Antwerp, Europe's second-largest port, to the North Sea. Ghent and Tournai fell to

the advancing Germans on 24 May. On 1 June, Hitler visited Ghent on a triumphal tour that included Brussels and Ypres. Also on the tour was the German War Cemetery in Langemark, where the commander of Hitler's World War I regiment is buried.

As planned, the BEF had moved forward to the Dyle river line at the start of the war. It couldn't stay there for long as the Germans forced their way through the Ardennes and towards the Channel coast south of the BEF. From 16 May the British retired in nightly bounds to prepared lines: on the night of 16/17 May to the Senne Line (Charleroi–Brussels–Willebroeck Canal); on 17/18 to the Dendre Line (Maubeuge–Mons–Ath–River Dendre to Termonde); and finally on 18/19 to the Escaut Line (the frontier defences to Maulde – the line of River Escaut to Ghent – thence the canal to Terneuzen). Avelgem (in the British war diaries Avelghem) is on the Escaut (called the Scheldt in Dutch) between Ghent and Tournai. There, 10th Brigade – elements of the East Surrey, Beds & Herts and Duke of Cornwall's Light Infantry Regiments – held the line until 22 May, when a further retreat, to the Yser, took place. This is an Artillerie-Regiment 50 sFH 18 near Avelgem, on 19 May. Note the matting under the gun's wheels.

Above: A Renault AGR petrol tanker in Wavre being inspected by a soldier from Artillerie-Regiment 50. The two great tank battles – the largest ever to that point – both ended in French success. In spite of their problems – mainly caused by the lack of radios in all the tanks – French armour inflicted a number of reverses on the Germans at Hannut and Gembloux, with the French artillery playing an important role. German tank losses were higher than the French, but because the French were forced to retreat, the Germans were able to recover and repair many of their losses. Tactical success, however, turned into strategic defeat as retreat was forced on the French by the success of the German thrust through the Ardennes. The attack through Belgium had done exactly what was asked of it: held the Allies' best troops away from the where the German Panzers were punching through weaker defences. French 1re Armée retreated to Lille, where its battle with the Germans gave the BEF long enough to evacuate through Dunkirk.

Below: A Luftwaffe Flak crew having fun in a garden somewhere near Bruges, watched by a rather bemused local. They wear the standard *Fliegerbluse* (flight blouse) with red collar tabs (*Kragenspiegel*) identifying them as Flak crew, each tab with one or more metal alloy cyphers called *Möwen* – gulls (picture gull wings in flight) – that indicate rank. Two of the men don't have the national emblem on their right breast, which was not uncommon in the early days. Two do – the man at right has a trade badge on his left sleeve as well; the man at left has an SA Sports Badge on his left breast pocket. Note there are no uniform buttons visible: to avoid snagging, the *Fliegerblusen* have no visible buttons.

Bridging stretches of water – lakes, rivers, canals – is an important requirement for an attacking army and essential for the Germans making their way across Belgium and the Netherlands, countries known for their waterways. This is an 8-tonne double ferry that is powered by two motors. A Brückenkolonne B (bridging column) could make a range of bridges or ferries from its equipment: 130m (142yd) of 4-tonne to 50m (55yd) of 20-tonne bridges; or eight 4-tonne, four 8-tonne or one 20-tonne ferries.

Belgium surrendered on 28 May and the Germans moved in. This photograph shows the arrival in Brussels of Feld Kommandantur (Field Command) 538. Subordinated to the Military Commander Belgium/Northern France, Field Command 538 was assigned to Korück 520 (Kommandant des rückwärtigen Armeegebiets, Commander of the Rear Army Area, later Oberfeldkommandantur 520) and oversaw the administration of the occupied territories and the German occupation troops until the end of 1940 when it moved to Poland. Here, Mercedes-Benz 230s towing 37mm Pak 35/36 anti-tank guns pass the Brasserie de Bruxelles.

BRASSERIE D E BRUXELLES
MONS-MOTO-CLUB

German troops in Brussels' Grand-Place. The grand houses are guildhalls, from right to left that of bakers (also known as 'The King of Spain'), greasers ('The Wheelbarrow') and carpenters ('The Bag').

The Netherlands

The German advance into the Netherlands was not quite the procession that soundbite online histories suggest. There were various defensive lines that included 2,000 pillboxes and the option to use flooding to defend large areas of the country. However, the Koninklijke Landmacht (Royal Netherlands Army) lacked manpower and the men in it were without sufficient training. It was seriously wanting in armour and sufficient air defences to stop the Luftwaffe bombers. What aircraft it had were attacked on the ground on 10 May and over half were destroyed. However, the Dutch gave a good account of themselves against the *Fallschirmjäger* who dropped on The Hague. Nor were they taken in by the many attempts by the German special forces, often dressed in Dutch uniforms, to take critical transport bridges. They put up a stubborn defence – so much so that Hitler's Directive No. 11 of 14 May said, 'On the northern flank the Dutch army has shown greater power of resistance than was expected. Political as well as military considerations require that this resistance be speedily broken.' The result was the destruction of the inner city of Rotterdam by carpet-bombing and the deaths of nearly 1,000 civilians. When Utrecht was offered the alternatives of surrender or to suffer the same fate, it became clear that the Germans intended to use the threat of aerial bombardment against civilians until the Dutch capitulated. The commander of the army, General Henri Winkelman, was left with no sensible option.

The German Army made great use of motorcycles in the west: in the east the terrain and road conditions made them less of a success. During the 1930s Germany overtook Britain as the largest user and manufacturer of motorcycles and continued their use into war. They were particularly utilised by motorised divisions. The *Kradschützen* battalion of a Panzer division's reconnaissance regiment had 250 of them and

three motorcycle rifle companies. The German 1939/40 mobilisation plan allocated 1,269 to each Panzer division. A motorised reconnaissance battalion had around fifty. These motorcycle troops – Zundapp KS750s with sidecar – advancing through the Netherlands are from 207. Infanterie-Division.

207. Infanterie-Division was established in 1939 as a border security unit during the invasion of Poland under Heeresgruppe Nord. In 1940 it was the main attacking force of 18. Armee against the Grebbe Line. With a history dating back to the mid-eighteenth century, the defensive line had been upgraded during the interwar years along French principles and was tactically out of date. Additionally, fire lines hadn't been cleared and the pumping station that would have allowed the ground in front of the line to be inundated hadn't been completed. 207. Infanterie-Division expected to overrun the line in a day; it took them three. The photo shows a horse-drawn 7.5cm Feldkanone 16 nA (*neuer Art* because they were old FK 16s with the original 7.7cm tube rebarrelled to 7.5cm) artillery unit passing a captured Dutch staff car. Note the seats for the two men on the front of the gun and the wooden wheels.

The Ju 87 earned itself a terrifying reputation during the land and sea battles of 1939–40, although it proved too vulnerable to RAF fighters during the Battle of Britain and had to be withdrawn because of high losses. Over the inexperienced Dutch Army, with almost complete air superiority, it was very effective and on more than one occasion Stuka attacks led to troops running away, including those defending the Grebbe Line.

France

Above: Map-reading is, of course, an important part of any soldier's skill. Before the war German military maps came from the 9th Department, Army Surveying and Military Geography. It was responsible for military mapping, surveying and geographical analysis. The Heeresplankammer (Army Planning Chamber) was responsible for the actual technical production. During the war the department was named Abteilung für Kriegskarten- und Vermessungswesen – Directorate for War Maps and Surveying. Getting the Panzers through the Ardennes would have required decent mapping.

Opposite above: The 3. SS-Division *Totenkopf* drew the majority of its men from concentration camp guards. Originally in reserve, from 19 May it supported Rommel's 7. Panzer-Division. In doing so it committed a range of war crimes against civilians, Africans in the French forces and PoWs in general. One egregious example was the murder of ninety-seven members of 2nd Battalion, Royal Norfolk Regiment. They would continue to commit atrocities right up till war's end, when they surrendered to US Army troops on 9 May 1945 in Czechoslovakia. The Americans, sensibly, handed them back to the Soviets. Note the SS uniform-identifying features: national symbol on left arm rather than right breast; the SS skull on the *Kragenspiegel* (the Panzer, Danziger version had no jaw) and skull on crusher cap (*Offizierfeldmütze älterer Art*) with leather peak. His pink *Waffenfarbe* colours identify a Panzer unit. He's in the turret of an SdKfz 222 *leichter Panzerspähwagen* armoured car armed with a 20mm cannon.

Opposite below: River crossings are always difficult and there are a lot of them in France. The first thing to do before any sort of bridge can be erected is to gain a bridgehead on the enemy bank. That usually means crossing in inflatable boats, sometimes powered, sometimes just with oars. Various things can help: surprise, artillery support (sometimes supplied by tanks), nightfall or smoke. The *Nebeltruppen* were trained in smoke and gas operations for just this reason.

The failure of the Germans to stop the evacuation of British and French troops from Dunkirk is one of the most discussed aspects of the Blitzkrieg period. There are many theories put forward for the reason the Panzers weren't allowed to complete the rout they had started. Nervousness seems to figure highly: the concern in the OKW that things had simply gone too well and a counterattack against an exposed flank was just around the corner. The promise by Luftwaffe commander Goering that the air force would handle it relieved the pressure but – as with almost every promise Goering made – the air force couldn't handle it. This KG 51 Heinkel He 111 is returning from a bombing trip over the Dunkirk coast.

Inset: Logistics are critical to any attack and the speed of the German advances in the first year of the war tested their systems to the extreme. With so much of the German Army horse-drawn, ensuring that far-ranging Panzer units had sufficient fuel and ammunition was a job for lorries and, occasionally, aircraft. In the west the Germans had sufficient motor vehicles. In the Soviet Union in later 1941 and 1942, the problems intensified. Here vehicles and horse-drawn trains pick up artillery ammunition from an ammunition dump. Note the wickerwork shell containers.

Opposite: A large number of British and French soldiers may have escaped the Germans, but they left behind their vehicles, heavy weapons and other equipment. Many of these were recoverable and the Germans had a field day. Each of the types taken into German service was given a title. The Universal Carrier, for example, was identified as the Bren 731(e) (where e = *englisch*) or Gepanzerter Maschinengewehrträger Bren (armoured machine-gun carrier Bren). Most were used without modification, the ammunition for the Bren light machine guns and Boys anti-tank rifles being available from captured stocks. However, there were also conversions: the 4cm Pak 192(e) which mounted a British 2-pdr, the 4.7cm Pak 188(h) with the Dutch Böhler 47mm, and the 2.5cm Pak 112(f) using a French 25mm anti-tank gun. A later modification was a *Panzerjäger* version which carried three 8.8cm *Panzerschreck* bazookas.

Below: A British Bedford burns outside Dunkirk. Many of the vehicles abandoned at Dunkirk and in northern France weren't destroyed. This treasure trove made a huge difference to the availability of lorries to support the German Army in the Soviet Union. There were problems, however: the variety of different vehicle types made spare parts – always a German Achilles' heel – and maintenance an even bigger problem.

Above: The Bussing-NAG G31 was produced in 1931–35 but was still in service in 1940, transporting troops or supplies.

Inset: The DKW SB 500 is not the best-known of the Wehrmacht's motorcycles but was used extensively in the German Army, here for *Kradmelder* (dispatch riders). Note the rubberised Kradmantel overcoats, which could be buttoned around the legs.

A 23. Infanterie-Division motorised column takes a breather. Having fought through the Ardennes, the division crossed the Aisne during *Fall Rot* and chased the French through Champagne. The vehicles are *leichte Geländegängiger Einheits Personenkraftwagen* (light all-terrain passenger vehicle) Kfz 1 – almost always known as *Kübelwagen* because of their bucket seats – built by the Stettin-based firm Stoewer. Two other companies – BMW and Hanomag – also made them, the main difference between the vehicles being the engines. Some 17,500 were built and they served throughout the war in four versions: Kfz 1 for general use, Kfz 2 for signals or maintenance equipment, Kfz 3 survey equipment and the Kfz 4 – the *Truppenluftschutzkraftwagen* (troop air protection vehicle) that had twin MGs on a double mount in the rear.

Inset: An excellent photograph that typifies the German Army of 1940, mixing the traditional – horse-drawn carts, infantry on foot – with the modern: MG 34s, sheet-metal carts, steel disc wheels and rubber tyres. The machine-gun carriage, Maschinengewehrwagen 36 (If 5), was part of the equipment for machine-gun companies of infantry battalions. The rear cart had a Zwillingssockel 36 pedestal mount for two machine guns (only one is attached in the photograph). One of the two gunners could ride in the mount ready for immediate action. He could detach the rear cart from the front and extend two supports to stabilise the gun cart. Increasingly from 1941 in light (or *Jäger*) infantry divisions, the double Maschinengewehrwagen 36 (If 5) was replaced by two single-horse If 8 infantry carts.

Three infantrymen at rest. Note the white for infantry *Waffenfarbe* on the soutache (inverted chevron) of their *Feldmützen*, the grey waterproof bag that contains a one-time use gas cape, the ubiquitous Kar98k bolt-action rifle and their distinctive belt buckles with the motto *Gott Mit Uns*, associated with the German military as far back as the Teutonic Order.

A bugler calls the unit to march again. While use of the bugle was rare in combat, its role was embedded into the army through long history and was still part of daily life.

A mixed infantry group on the road in France. Note the 3.7cm anti-tank gun, the development of which started back in 1925 as a *Tankabwehrkanone*, although the terminology later changed to *Panzerabwehrkanone*. When conceived, and well into the 1930s, the 3.7cm Pak was a match for any AFV. Over 11,000 were available in 1939 and it was the standard German infantry anti-tank gun of Blitzkrieg. It was during the French campaign, when heavier tanks were encountered for the first time, that its lack of penetration made itself felt, and as the war went on, it became known as the *Panzer-Anklopf-Kanone* – the 'Doorknocker'.

Inset: Break step over bridges! The command has been given since the 1831 Broughton Suspension Bridge collapsed as soldiers marching in time crossed it. In fact, the problem only really occurs when the marching frequency matches the mechanical frequency of the bridge – which is rare. Either way, the soldiers in this unit crossing a pontoon bridge don't seem to be worried. Note how much of an overview is afforded to the officer on horseback in the middle of the bridge.

A weary unit prepares to march on. Note the torches clipped to the front of three NCOs' *Koppeltragegestell mit Hilfstrageriemen* (belt carrying frame with auxiliary carrying straps) – the Y-straps introduced in April 1939 that form the basis for carrying equipment and distributing the load evenly. From right to left:

Oberfeldwebel (binoculars and DRS Sports Badge), *Feldwebel* (Kar98k and ammo pouches on belt) and *Unteroffizier* (machine pistol 38 or 40 without magazine and canvas magazine pouches). Note that full pack and greatcoats are being carried.

Another pontoon bridge, this one over the Meuse at Monthermé, reached by Generalleutnant Hans Reinhardt's XLI. Panzerkorps on 13 May. The first two attempts to cross by rubber boat failed, but the third succeeded, as did another further upstream where a bridge hadn't been completely destroyed. The heights were difficult to take but Pionier-Bataillon 57 erected a 20-tonne pontoon bridge quickly under fire and the Panzers were able to cross and push forward. Once the enemy had been routed, the 6. and 8. Panzer-Divisionen advanced quickly and by nightfall on 15 May they were nearly 65km (40 miles) deep into the French lines.

While the devastation caused by the fighting in the west in 1940 came nowhere close to that of 1944–45, nevertheless there was a heavy civilian price paid by the French and serious infrastructure damage. By the end of the war more than half a million French civilians had died, many through bombing by both the Germans and Allies.

Between 12 and 17 May, on the banks of the Meuse, the Battle of Sedan saw the German strategic plan for *Fall Gelb* come to fruition. With fighting raging in Norway and the Low Countries, the eyes of the Allies were fixed on what they thought was the main German thrust: the attack by General Fedor von Bock's Heeresgruppe B. Relatively undefended because the French High Command had believed the Ardennes impenetrable – in spite of much evidence to the contrary – Guderian's XIX. Armeekorps captured Sedan and crossed the Meuse quickly, supported by immense aerial activity from Luftflotten 2 and 3. The critical day was 13 May, but the French missed their chance for a swift counterattack. By the time they did, sufficient German forces had crossed the river. The Armée de l'Air and RAF attacked the bridgeheads in strength but they were decisively beaten and the Germans were able to push 600 tanks across the Meuse by nightfall on 14 May. Guderian – countermanding the instructions from the OKW to wait for his infantry divisions who were still struggling through the Ardennes – pushed two of his three Panzer divisions – 1. and 2. – on west towards the French coast. Held in reserve during the attack on Poland, 10. Panzer-Division had joined Guderian's XIX. Armeekorps for *Fall Gelb*. He tasked it and the *Grossdeutschland* Regiment to hold Sedan. Between 15 and 17 May, these troops fought off the French counterattack at Stonne. The village changed hands seventeen times, was the scene of Char B1 bis *Eure*'s epic attack (see p.179), but was ultimately held and reinforced as the German infantry arrived on the scene. This photograph shows one of 10. Panzer's thirty-two PzKpfw IVs, this one from 4./Panzerregiment 7.

The Luftwaffe had been dominant over Poland but had a more difficult time in the west. The ground-support element provided material assistance to the German thrust through the Ardennes and across the Meuse, quickly moving to forward airfields so that they didn't let their ground forces get out of range. On 17 May they were operating out of Charleville, resupplied with fuel, ammunition, spare parts and groundcrew by Ju 52s. However, over Dunkirk, it was a different story. The Bf 109s were still further from the area than the British fighters based in southern England, so the airspace above the evacuation area was well contested: in the fighting from 26 May to 4 June, the RAF lost 177 aircraft; the Germans lost 240. Murray cites Fliegerkorps II's war diary reporting that it lost more aircraft on the 27th attacking the evacuation than it had lost in the previous ten days of the campaign. This attrition and the hard fighting played a part in the failure of the Luftwaffe during the Battle of Britain: the Luftwaffe was in continuous action from May until the end of September 1940, had lost over 30 per cent of its aircraft over France and was in need of a rest and recuperation period. Accidents didn't help as this image of a crashed Bf 109E exemplifies. The 109 killed a lot of pilots because it had narrow landing gear. Putting down on a muddy field, this one most likely flipped and broke its back possibly killing the pilot. The Bf 109 was a hugely successful design responsible for more aerial kills than any other aircraft. The 'Emil' version was built in large numbers from 1938 – around 3,500 – and it performed well against French aircraft and British Hurricanes over France, although the Spitfire was at least its equal.

An important interwar aircraft, by 1940 the Potez 25 was used for training, second-line and observation/reconnaissance duties – the latter 25 A2 version, as here, with two seats. By the time the campaign ended with the signing of an armistice on 22 June, the Luftwaffe had lost 1,469 aircraft on operations (a massive 20.9 per cent through accidents), the French 892 and the British 1,029.

Charleville-Mézières, west of Sedan, fell on 14 May. This is the beautiful Place Ducale, which sustained some damage in the attack. The German forces, spearheaded by Hoth's XV. Armeekorps (which included Rommel's 7. Panzer-Division) flooded through the Sedan gap, heading for the Channel coast.

33. Infanterie-Division support troops follow the advance along the route de Charleville in Maubert-Fontaine, heading towards Saint-Quentin. In all the excitement of the Panzer thrust to the coast it's easy to forget what was happening behind the spearhead: the infantry rushing to keep in touch; the support elements – from *Gulaschkanonen* to fuel reserves to baggage trains; and the follow-up troops. Here, two columns – one horse-drawn (main photograph), one motorised (inset) – of medical units from Sanitäts-Dienste (Medical Services) 33, which were made up of two *Sanitäts-Kompanien* (medical companies), two *Krankenkraftwagen-Züge* (ambulance platoons) and a *Feldlazaret* (field hospital). The horse-drawn column includes two Hf 11 or 13 *grossen Feldküchen* (large field kitchens).

KISS
N.39
HIRSON

World War I Maschinengewehr 08 saw service as a heavy machine gun in many infantry divisions in 1939–40. Heavy, water-cooled, with a sledge mount (*Schlittenlafette*), it was relegated to second-line service as the MG 34 and 42 became more available.

A 15cm sFH 18 battery ready for action. Heavy artillery was usually some way back from the front line – sometimes as much as 9–10km (5–6 miles)– in a camouflaged position such as on the edge of a wood as here. This meant that if they were horse-drawn – as so many were – the horses could be kept under cover in the wood. Each gun has ready ammunition (the projectiles are visible at left next to the trailer wheels); there will be a main ammo dump under cover. Note the profusion of camouflaged tents. The position will have trenches for the guncrews in case of counterbattery fire or air attack.

Firing a 15cm sFH 18. Most of the seven visible guncrew have their hands over their ears. The number of guncrew varied from five (the minimum required) to twelve (needed if horse-drawn to assemble and disassemble). Note the wickerwork projectile containers and the wooden boxes holding the charges. One of the four battalions in each infantry division's artillery regiment had these howitzers. The use of an SdKfz 7 to tow the weapon speeded up movement (towing speed by horse was at least five times slower) and also the time taken to split the weapon into two loads for transportation by horse. When towed by the SdKfz 7 in one piece it took half the time to ready for action and half the manpower.

Above: Sporting a typical Wehrmacht officer's grey pennant on his Mercedes-Benz 320 (W142) cabriolet staff car, the driver of 4. Panzer-Division's commander smiles for the camera. He's wearing an SA Sports Badge. The commander of the division from 6 April 1940 was Generalmajor Johann Stever, who took over from Generalmajor Ludwig Ritter von Radlmeier, who had fallen ill. Stever was wounded on 14 May at the Battle of Hannut, and for five days Oberst Hans Freiherr von Boineburg-Lengsfeld took over. Stever took up the reins again on 19 May and remained in position until the end of July, when von Boineburg-Lengsfeld took over.

Burnt-out German reconnaissance vehicles: a *Panzerspähwagen* (SdKfz 231) eight-wheeler and an SdKfz 232 six-wheeler (at right) with its distinctive bedstead antenna. Both were armed with a 2cm KwK 30/38 L/55 autocannon and an MG 13 or MG 34. Many more of the eight-wheelers (about a thousand in total) were produced. They replaced the six-wheelers (about 130 produced) on the production lines. Panzer divisions relied on the information supplied by the *Aufklärungsabteilungen* (reconnaissance units), who were equipped with armoured cars and SdKfz 250 or 251 halftracks for infantry sections.

Below: Friendly fire (today we call them blue-on-blue incidents) has always been a common feature of war. As the lethality of weapons systems improved, so IFF – identification friend or foe – systems became more important. In its most basic form this is about insignia – crests on shields, swastika flags on turret tops – and communications, or their lack (probably the most common cause). Hitler was aware of the problems: on 26 February he sent out a directive to ensure that 'everything humanly possible be done to prevent mistaking our own troops or troop units for those of the enemy'. Of course, it didn't always work, as this photo shows. Chémery-sur-Bar is just south of Sedan. On 14 May a 1. Panzer-Division column was attacked by Stukas and there were important casualties: the staff of Panzerregiment 2, a PzKpfw II and III, and this unusual 8.8cm Flak 18 mounted on an SdKfz 8 – one of only ten made.

Above: The Char B1 bis was a powerful and sophisticated vehicle armed with a 47mm SA 35 or 34 anti-tank gun and 75mm infantry support weapon. When they worked – and they were notoriously unreliable – the B1 and B1 bis were as good as anything on the battlefield. The best example of this were the exploits at Stonne, 25km (15 miles) south of Sedan, of No. 337 *Eure*, the B1 bis commanded by Capitaine Pierre Billotte of the 41e BCC, 3e DCR. On 16 May he destroyed thirteen German tanks – two PzKpfw IVs and eleven PzKpfw IIIs. *Eure* was reportedly hit 140 times but not penetrated. In the end, Billotte despatched his own vehicle on 13 June so it didn't fall into the hands of the enemy. Around 128 B1 bis were lost in combat; 139 abandoned or destroyed by their own crews due to breakdowns or fuel shortages.

Famous for the significant use of tanks in the nearby battle during World War I, on 18 May Cambrai was taken with ease by Rommel's 7. Panzer-Division after it had been bombed by the Luftwaffe the day before. From Cambrai, 7. Panzer-Division advanced towards Arras, where British 4th and 7th Royal Tank Regiments counterattacked and struck the Panzer division's exposed flank. However, as with Cambrai in 1917, the initial British success was quickly negated – in 1940 by the arrival of a number of 8.8cm Flak 36 guns and Fliegerkorps I's Stukas. Nevertheless, the attack sowed the seeds of doubt in the minds of Hitler and OKW and the Panzers' move towards Dunkirk was halted at a critical moment.

Opposite below: The Somua S35 was produced from 1935 to 1940, the French Army having about 350 in 1939. It was expensive to make, had a one-man turret with a hatchless cupola – meaning the commander had to fight from inside, thus reducing his tactical awareness because of poor visibility – was automotively unreliable, and there were too few radios for each vehicle to have one. However, it had a good gun, a 47mm, and good armour. In the battles of Hannut and Gembloux it was only bettered by the PzKpfw IIIs and IVs – and then mainly because of the lack of radios. This one was knocked out by 4. Panzer-Division. The heart identifies 2e peleton (Section): 1er = spades; 3e = diamonds; 4e = clubs). Its colour (red) signifies 3e escadron (Company): blue = 1er; white = 2e.

Two photographs of abandoned Hotchkiss H 39s (or to give them their full name: Char léger modèle 1935 H modifié 39). This was not the most successful of tank designs for a variety of reasons. It had a crew of two, which made life difficult for the commander. To make matters worse, the commander had no hatch in the cupola, which meant he spent much of his time sitting on the large escape hatch a long way from the main gun – a semi-automatic 37mm Model 18 (SA 18). Automotively, the tank was unreliable and had very poor cross-country handling – bad enough for the French infantry to refuse it. Politics (and the fact that a lot had already been produced) meant that it went into service with the cavalry (300), the infantry only taking the last 100. The improved APX-R turret allowed for the upgunning with the longer 37mm SA 38, and the H 39 was a major improvement with a more powerful motor. Around 700 of the H 39 were produced. The Germans captured about 550 and used them with a new cupola as the PzKpfw 35H 734(f) and 38H 735(f). The two here are both H 39s, one (above left) with the longer SA 38 gun and a diamond ace indicating 3e peleton. The other (above right) has the shorter SA 18 and ace of clubs (4e peleton).

The fighting led to refugees, a high percentage being women and children trying to find a safe exit from the war zone. These homeless, dispossessed people ranged from those who had left their homes with cartloads of possessions to those who had only what they stood up in. The French call it l'exode (the exodus) and it saw people from the Low Countries mixing with northern French heading south. Soon they were joined by Parisians: three-quarters of the urban population left – there were 200,000 hiding in the forest of Fontainebleau. After the exodus, there was the great return: 3 million by road or rail, before 1 November 1940. A note from the Refugee Directorate dated 7 March 1941 to the German Embassy estimated the number of French refugees at 7 million, of whom just under a million had not yet been able to return home. Many of them returned by train: records show that in the period from 21 to 31 August, the operation of 375 trains, including eighty-six trains for demobilised personnel, seventy-nine refugee trains for the zones between the Loire and the Somme, forty-four for Alsace-Lorraine and sixty Belgian refugee trains.

The refugees were at first civilians; later, increasingly, servicemen who had become separated from their units or fled the battlefield – clogging roads and other transport facilities, making defensive coordination a problem. This exodus was compounded by the population of other countries – in particular, 2 million Belgians, alongside people from the Netherlands and Luxembourg – who fled from the advancing Nazis. By the time the German Army reached Paris, which had been declared an open city to reduce infrastructure damage, as many as 10 million displaced persons had fled, almost a quarter of the French population.

There aren't many guards on this column of PoWs; there were too few to go around because there were so many PoWs to look after, about 1.8 million. Many were held until 1945 in German PoW camps and over 93 per cent of them were made to work in Germany – in agriculture, industry and mining. Some 37,000 died in captivity.

The French corps of *tirailleurs sénégalais* (Senegalese riflemen) was created by Napoleon III in 1857. Their uniform was distinctive for its famous red cap or *chéchia*. This one was probably donned for the photographer. Used in parade or walking-out dress, the *chéchia* were sometimes worn under the helmet or with a khaki cover. The Black French troops suffered many atrocities when captured by the Germans in 1940. Many were shot; some were tortured. Unsurprisingly, they became notorious for their uncompromising behaviour later in the war.

This motorcycle messenger looks dusty and tired, as so many soldiers were. There were few pauses in the advance into France. The Wehrmacht had an answer for this: Pervitin, a methamphetamine, is the best known. As Nicholas Rasmussen put it: 'the German Blitzkrieg was powered by amphetamine as much as it was powered by machine.'

Head lice are always a problem for soldiers – more so in trench warfare or confined spaces – and that's one good reason for a short back and sides.

Fires and smoke aren't always a good idea when in enemy territory, but if there isn't a *Gulaschkanone* around …

German soldiers in a French orchard. They are from a weapon collection detail, picking up anything useful from the battlefield.

TRANSPORTS CITROËN
ARRÊT
FAIRE SIGNE AU
CONDUCTEUR

Coffee breaks could always be supplemented by whatever was available locally. Sometimes 'requisitioned', sometimes simply purloined – troops spending time in a locality, particularly in the west, were more likely to behave properly as ordered by regulations.

Hot food is important for troops in combat and the main German preparations were made courtesy of the *Gulaschkanonen* – the mobile field kitchens. The *grossen Feldküchen* – often on the backs of lorries – were capable of feeding around 200 men utilising a 200-litre (44-gallon) cooking vessel, a 90-litre (20-gallon) coffee kettle and a 70-litre (15-gallon) roasting and braising oven. Each had a crew of three, although the quality of the cooking depended on the skill – or otherwise – of the cook.

A welcome coffee from the *Gulaschkanone* (goulash cannon, so called it is said because of its long—here extended—chimney) during the advance into France. The German Army's food supply relied on a combination of rations cooked by the field kitchen and that picked up locally. Each soldier carried a day's

worth of iron rations but these could only be opened with permission—usually if there was no opportunity to get field rations. Most field kitchens were towed by horses—as here—and weren't always able to keep up with the troops.

Nearly 2,500 of the Krupp L3 H163 3.5-tonne *mittlerer geländegängiger Lastkraftwagen* (middle-weight, cross-country lorries) were built. A major version, the Kfz 72 had a closed body with vertical wooden slat walls and was used for radio, telex, telephone, radio listening, weather and printing vehicles. This is a Kfz 72 *Funkkraftwagen* (radio lorry) a or b with a 10m (33ft) *Steckmast* (radio mast) that was stored in sections on the left side of the lorry.

After completing its formation, 94. Infanterie-Division moved to the Saarpfalz region to secure the German western border. At the beginning of the western campaign, the division served as an OKH reserve and did not participate in the fighting in France until 27 May. It took part in the fighting on the Escaut (the Scheldt) south of Cambrai with 6. Armee. This photograph shows a signals unit from the division installing telephone lines. Most armies in World War II made heavy use of field telephones, radios being used mainly with mobile units or aircraft. The main German telephone was the FF 33, which had a 1.5V battery to power the handset microphone and a crank magneto to generate ringing current.

Wire-laying using a *Rückentrage für Feldkabel* (backpack dispenser for field cable). Each spool could hold approximately 1km (0.6 mile) of single-conductor field telephone wire. Typically, the wire was dispensed from the spool and suspended from trees or wire forks (*Drahtgabelen*). Sometimes, however, it just had to be laid on the ground. The trouble with this is that an enemy – if they get close enough – can eavesdrop on one-wire phone conversations through the ground.

Above: Army Red Cross vehicles during the invasion of France. Under the control of the Nazis the Deutsches Rotes Kreuz (DRK – German Red Cross) did whatever they were told to do and, while they cooperated with the International Committee of the Red Cross concerning PoWs, they supported the regime's policies and actively obstructed many attempts to relieve the suffering of concentration camp inmates. This stationary convoy of Wehrmacht medical vehicles awaits instructions somewhere in France. The leading vehicle is an Opel 3.6-47 type W39 omnibus built on the Opel Blitz chassis, which was put into series production at the end of 1939. Over 8,000 were built.

Opposite above:A roadside burial. The battle for France saw 90,000 French soldiers die and some 1.8 million enter captivity. The German Army's death toll was around 27,000 dead and 18,000 missing.

Opposite below: On 25 March 1940, Dr Leo Miller was appointed Catholic division chaplain of 57. Infanterie-Division, which fought its way through Luxembourg and southern Belgium and then from Saint-Quentin to Abbeville. He took this photograph of, and probably officiated at, a Luftwaffe military funeral. The air force lost 28 per cent of its strength – around 1,400 aircraft and 3,000 dead and missing – during the Battle of France. The Nazis had a very ambivalent attitude to the Church. On the one hand, there was a significant atheist element that was, at its most extreme, anti-Church. On the other, there was a large religious community in the army that wanted to see religious burials.

24-0519-F

7 *Fall Rot* – The Fall of France

WITH THE BULK OF the British troops – and some 170,000 French – extracted from the beaches at Dunkirk, the Germans had missed a chance to win an even more devastating victory than they already had. The Panzers had been stopped from advancing after the British counterattack at Arras had unnerved the high command. Protect the flanks and wait for the infantry – these were the watchwords coming from the Supreme Commander. On 23 May the British thought Dunkirk would fall on the 24th; then came another halt order. The tanks waited because the Luftwaffe was going to be responsible for the destruction of the port. Hitler wanted to hold the tanks back for *Fall Rot* – the next phase of the attack into France – and certainly didn't want any chance of blue-on-blue accidents, something that was one of his particular bugbears (see p. 178).

The ground attack resumed on the 27th, but by that time most of the troops had got away. On top of this, the defenders fought tenaciously, the terrain wasn't the best for tanks and – most importantly – rain on the 26th and 27th had made the ground less suitable for armour and grounded the Luftwaffe. Even Guderian said, 'The infantry forces … are more suitable than the tanks for fighting in this type of country and the task of closing the gap on the coast can therefore be left to them.'

The tenacity of the defenders of Dunkirk was to be repeated in the next weeks as the Germans sought to extend their domination over the whole of France. First, however, they replenished their forces, gave their exhausted troops some respite and positioned themselves for the final conquest. Ranged against them along the line of the Somme–Oise–Ailette were sixty French and four British divisions. The second BEF, as it became known after the war, was made up of a polyglot grouping of troops: Beauman Division, mainly of reservists, logistics and communications troops, and pioneers who had been between the BEF and the main British bases (the ports of Brest, Cherbourg, Dieppe, Le Havre, Nantes, Rouen and Saint-Nazaire); 51st (Highland) Division that had been in the east of France and joined French 10e Armée south of the Somme; and the 1st Armoured Division that had landed in France between 15 and 26 May. Lieutenant General Alan Brooke was sent over to coordinate the action along with reinforcements from Britain: a 52nd (Lowland) Division infantry brigade group and Canadian troops from First (Canadian) Army which headed for France on 7 June.

It was to no avail. The advance of Heeresgruppe B down the west coast of France was unstoppable. 51st Highland Division surrendered at St Valéry-en-Caux on 12 June, the day before Paris was declared an open city. However, many of the remainder of the British troops were able to retreat and return home from the west coast ports before these were taken by the Germans; Operations *Cycle* (from Le Havre) and *Aerial* (mainly Nantes and Saint-Nazaire).

The momentum of the German forces pushed them on in spite of the best efforts of the French forces, whose fighting troops did what they could. Martin Alexander gave an example:

> The 29th Alpine Division's 'shooting war' did not begin till 5 June. Yet by 25 June it had lost 70% of its field officers and 65% of its subalterns. Of the division's constituent units, the 6th Chasseurs alpins demi-brigade lost 72% of its officers defending the strong points south of Peronne between dawn on 5 June and nightfall on 6 June. … By the armistice its officer casualties totalled 93% (95% in 25th Chasseurs and 75% in 65th).

Italy opportunistically declared war on France on 10 June. Its attacks saw little success but added to the political pressure on the French cabinet. The armistice came on 22 June. About 1.6 million French troops became PoWs, but as Alexander put it: 'The armistice put them into German captivity; it was emphatically not that their capture caused the armistice.' The Germans suffered c.157,000 casualties in the fighting in the west, including over 45,000 killed or missing. More than half of the casualties came during *Fall Rot*.

The armistice was signed on 22 June and came into effect on 25 June, the same day as that with Italy. Cleverly, the Germans left a French government in place so that the French Navy and its colonial possessions wouldn't continue the war. Vichy France – the southern part of France less the west coast – remained unoccupied until November 1942 when Operation *Torch* triggered *Unternehmen Anton*.

A development of the Char NC, the Char D1 was longer, had a wider turret ring to take a bigger turret armed with an SA 34 47mm main gun, had heavier armour and was fully equipped with ER 52 and ER 52 bis radios, the aerial for which can be seen jutting out of the back at an angle of 45 degrees. It has a typical French trench-crossing tail. The D1s were poorly put together and so were shipped out for use in the French colonies. They were brought back to France in extremis in May 1940. Around forty came back and their first major engagement was against 9. Panzer-Division, where many were destroyed and the rest captured. They became the PzKpfw 732(f) in German use.

The precursor to *Fall Rot* was an Anglo-French counterattack in the Abbeville area on 27 May. France was divided along a line from the Maginot fortifications in the east along the rivers Aisne and Somme, ending at the mouth of the latter, Saint-Valery-sur-Somme in the west, although the Germans had established a number of bridgeheads over the river. Reinforced by the arrival of British 1st Armoured Division from England and 51st Highland Division from the Saar region, the Allies attacked the Germans. There was a moment in the attack, in the early hours of the 29th, when it might have succeeded had it pushed on, but – with the situation unclear – the Allies stopped to regroup, the Germans reinforced and the 8.8cm Flak guns used in the ground role took a deadly toll on French armour. The offensive ended on 30 May with a small German bridgehead holding on and the French tank losses reaching over 100. The photographs show the aftermath of the battle: damaged Allied aircraft (a Lysander in the foreground), French Canon CA de 75 modèle 1932 AA guns; and M884, a Somua S35 of 29e régiment de dragons (Dragoon Regiment).

SOMUA
M 884

Above: A knocked-out SdKfz 222 armoured car outside Porte Saint Paul, Verdun. The statue is by Auguste Rodin and was originally produced as Rodin's entry to a competition for a monument commemorating the 1870 defence of Paris. It was chosen in 1916 by the Dutch committee of the League of Neutral States as a gift to Verdun inscribed:

A la France Immortelle
a la glorieuse ville
de Verdun
la Hollande Amie

(To immortal France
To the glorious city
of Verdun
Friendly Holland)

Today the statue of a wounded warrior supported by winged Victory can be found on the banks of the Meuse on rue des Frères Boulhaut. While Verdun held out for nine months in 1916, in 1940 it was captured on 15 June. Two days later, the 'hero of Verdun', Maréchal Philippe Pétain, asked for an armistice. General Charles de Gaulle replied from England on BBC radio: 'Whatever happens, the flame of French resistance must not and will not be extinguished!'

Below: *Fall Rot* began on 5 June after the Germans had mopped up the Channel coast around Boulogne, Calais and Dunkirk. Pushing towards Rouen, on 9 June they crossed the Seine and trapped 51st Highland Division and French 9e corps d'armée. Cut off from their supply lines, with their back to the sea, there was little these troops could do other than surrender. Four and a half years of imprisonment beckoned for most of these soldiers, although just under 3,500 were taken off the beaches. However, elsewhere on the west coast of France there were still large numbers of British and Allied troops in France, and a further 190,000 of them were rescued in Operation *Aerial*. In total, that meant over 550,000 saved during the Battle of France, of which nearly 370,000 were British.

Above: The PzKpfw IV was designed for infantry support and in 1940 there were 278 equipping ten *Panzer-Divisionen*. Poland had shown the Germans that the PzKpfw IV wasn't sufficiently well armoured but the uparmoured Ausf E was developed too late for *Fall Gelb*. The heaviness of the fighting saw ninety-seven PzKpfw IVs lost in the west. These are seen in Verdun.

Below: Also in Verdun, these PzKpfw IIs show a typical paint scheme: RAL 7021 dark grey. Note the outline turret lettering and the smaller tactical numbers on the rear.

The Maginot Line

The Maginot Line was a complex system of strongpoints constructed by the French against attack by Germany. It stretched from Belgium to Switzerland and from there to the Mediterranean. Total length was approximately 670km (416 miles), with the most heavily defended stretching about 450km (280 miles) along the Franco-German border. The soldier is examining a rotating armoured turret mounting a short-barrelled 75mm M1897 howitzer. The armour was typically around 35cm (13.8 inches) thick.

Preserved as a war memorial, the Ouvrage de la Ferté between Villy and Ferté-sur-Chiers was in the fortified sector of Montmédy. After the German attack through Sedan and the retreat of many of the troops in the area, it became the western edge of the Maginot Line. German 71. Infanterie-Division attacked what they called Panzerwerke 505 (Armoured Work 505) on 15 May and the fighting was intense. By 19 May there was no response from the defenders – Lt Maurice Bourguignon and 100 others – and the interior was on fire. They were all later found dead of carbon monoxide poisoning. The defenders were honoured with posthumous decorations, Bourguignon with the *Légion d'Honneur*, the others the *Ordre de l'Armée*.

Today, we think of the Maginot Line as being a white elephant. It cost a fortune to build but failed in its role of deterring an invasion. In reality, the story is more nuanced. The Maginot Line, it is true, didn't stop the invasion, but it forced it north away from France and into Belgium. It protected Alsace-Lorraine, which had been reunited with France by the Treaty of Versailles. It proved difficult to assault – most of the fortifications were unbowed, only ten of the main fifty-eight were captured – and the garrisons wanted to fight on. But surrender they did and by early July their garrisons had become PoWs. Photograph taken by men of 167. Infanterie-Division.

Block 6 of Ouvrage Hochwald was an impressive combat block boasting three 75mm gun embrasures, and turrets for a grenade launcher, automatic rifle and twin machine guns. Part of the fortified sector of Hagenau, it was designed to provide artillery support of the Hochwald region of north-east France. Operational from 1933, it saw action between 16 and 22 June 1940 and didn't fall to the Germans until after the armistice.

Rivers have their uses: allowing the refreshment and cleaning of men and horses after long, dusty days in the attack. Note the WH (= Wehrmacht) marking on this Opel Admiral Cabrio and the tactical marking on the Kfz 305 Opel Blitz ambulance. The petrol engine proved itself in the cold conditions of the Eastern Front when diesel lorries had problems starting. Picking up wounded from the *Wagenhalteplatz* (ambulance post), the 'Sankas' took them to either the *Hauptverbandsplatz* (field dressing station) or *Leichtverwundetensammelplatz* (collecting point for lightly wounded). The badly wounded were taken on to the *Feldlazarett* (divisional field hospital). The unit is Heeresgruppe C's 167. Infanterie-Division and the location is east of Metz, where the division was involved in capturing Ouvrage Kerfent and Bambesch of the Maginot Line on 20–21 June.

Four infantrymen in France, all armed with Kar98k rifles with ammunition pouches on their belts. Man at left has two Stielhandgranaten 24 in his belt and a Model 39 Eihandgranate 24 in his left hand. Man at far right also has two Model 24s.

B1 bis No. 539 *Drouot* of the 28e bataillon de chars de combat; alongside No. 30036, a char léger modèle 1936 FCM light infantry tank of the 4e BCC. On 16 June, in Saint-Germain des Prés (Loiret), *Drouot* and sister B1 bis *Galieni* were brewed up by two PzKpfw IIs. The rest of the unit retreated to Gien.

Variations on a theme: abandoned FTs. This (above) is at Verdun; the other's location is unknown. The French camouflage schemes were more flamboyant than those of the Germans. The *automitrailleuse à chenilles* (armoured car with tracks) Renault FT modèle 1917 was produced in 1917–19 and was obsolescent in 1940. In spite of this, the French Army had a number of tank battalions equipped with the FT – when war was declared, 1,445 FT tanks were in service on mainland France, and 1,235 unarmed. The Wehrmacht captured around 1,700, of which some 500 were reconditioned and used as tractors for things such as Luftwaffe aircraft, or for training. There are stories of the FT continuing in service in developing countries into the 1950s.

Gien, east of Orléans, was an important defensive location on the Line of the Loire with a strategic bridge that the Germans wanted to take. The town was held by 23e division d'infanterie and was heavily bombed, culminating in the raid on 15 June that set the town on fire. The resulting conflagration lasted three days and burnt over 700 buildings, including the famous fifteenth-century royal castle, whose towers can be seen in the background of this photograph. The French tried to blow the ancient bridge but the damage was quickly repaired by 1. Gebirgsjäger engineers, the division being three times the first across a major river: the Maas, Aisne and Loire. After the war, the town and castle were rebuilt and restored.

A sequence of photographs showing a 10.5 cm le FH 18 battery preparing for action. First (opposite above), a mounted messenger brings a fire request. The battery commander receives the request and passes out the fire order (below). The crew moves to the weapon and prepares to fire (opposite below).

Opposite: The French prime minister, Paul Reynaud, resigned on 16 June and President Lebrun invited the 'Hero of Verdun', Maréchal Philippe Pétain, to form a government. Immediately discussions about an armistice began and while they continued, the Panzers kept advancing. The tanks reached Brest on 19 June and the same day crossed the Loire at Nantes and Saumur. The furthest German advance was down the west coast, XV. Panzerkorps reaching Royan on 25 June and Saint-Jean-de-Luz near the Spanish border on the 17th. Here, German officers look over the old port of La Rochelle, 80km (50 miles) north of Royan, from the cours des Dames.

Below: There had to be a victorious arrival into Paris and it took place on 14 June. The Germans would then rub French noses in it by choosing Compiègne – where the 1918 Armistice had been signed – as the location for signing the 1940 armistice on 22 June. Paris had been declared an open city and there was little fighting. The four years of occupation that ensued saw the looting of the French and European economies as their financial institutions were raided, their resources taken, their industry diverted to German requirements and their manpower drafted for use in German factories.

8 The Aircraft of Blitzkrieg

On 1 September 1939 the Luftwaffe was the strongest air force in the world. Over the next two years it proved itself the perfect foil for the Panzers. Its powerful AA (anti-aircraft) units, the skill of its *Fallschirmjäger* and the transport units that dropped them by glider or aircraft, and the excellence of its bombers and fighters meant it had few rivals.

There were blots in its copybook. It wasn't able to subdue the RAF, which left the invasion of Britain in abeyance. Aircraft production was not as high as it needed to be. The number of transport aircraft, as an example, would be limited throughout the war. Additionally, development of next-generation aircraft foundered when the He 177 and Me 210 proved problematic. With too much time later in the war spent on research and investment on jets and V-weapons, by 1944 the Allies had the better aircraft.

The Germans realised the problem in 1942 but had little time to make changes. They increased production substantially but the aircraft they produced were little better than those used against Poland.

Directive No. 9 Principles for the Conduct of the War against the Enemy's Economy of 29 November 1939

1. In the war against the Western Powers, Great Britain is the driving spirit and the leading power of our enemies. The conquest of Britain is, therefore, the prerequisite for final victory.

The most effective means to achieve this is to paralyze Britain's economy through interrupting it at critical points.

2. … The necessary provisions must therefore be made as early as possible for striking a destructive blow at the British economy by concentrating suitable weapons of our Armed Forces on the most important targets. …

3. As soon as the Army has succeeded in defeating the Anglo-French field army and in occupying and holding a part of the coast facing England, the task of the Navy and the Air Force of carrying on warfare against the economic structure of Britain will become of prime importance. …

4. To the Navy and Air Force will fall the following joint tasks, enumerated here in the sequence of their importance:

a. Attacks on the main British ports of transshipment by mining and blocking the approaches to the harbours and by destroying vital port installations and sea locks.

In this connection the role of the mine-laying aircraft will be a very important one, especially with regard to the harbours on the west coast of Britain, in narrow waterways, and estuaries.

b. Attacks on British merchant shipping and against the enemy fleet protecting it.

c. Destruction of British supplies of oil and of food in cooling plants and grain elevators.

d. Interference with British troop and supply transports to the French coast.

e. Destruction of industrial plants, the elimination of which is of decisive importance for the conduct of the war; above all of key-plants of the aviation industry and the factories producing heavy ordnance, anti-aircraft guns, ammunition and explosives.

5. The most important transshipment ports in the British Isles, which handle 95% of the foreign trade and could not be adequately replaced by others, are:

London
Liverpool
Manchester

These three harbours, handling 58% of England's peacetime imports, are of decisive importance, especially for imports of food, timber and oil, and the processing thereof. For the export of coal, the following are important:

Newcastle	Swansea	Blyth	Cardiff
Blyth	Cardiff	Hull	

The following may be considered alternate harbours, but to a limited extent and for certain goods only:

Grangemouth	Holyhead	Leith	Bristol
Middlesbrough	Belfast	Grimsby	Newport
Southampton	Goole	Glasgow	Dundee

6. In harbours which cannot be blocked effectively with mines, merchant shipping is to be paralyzed by sinking ships in the roadstead and by destroying vital harbour installations. Special emphasis is to be laid upon the destruction of the great canal locks at the harbours of Leith, Sunderland, Hull, Grimsby, London, Manchester (Ship Canal), Liverpool, Cardiff, Swansea and Bristol-Avonmouth. Particularly on the west coast these locks are very important in regulating the water level and, through it, the harbour traffic.

There were other problems. First and foremost, German aircraft losses were high – as were losses of aircrew – and these were not just as a result of enemy action. Fully 750 of the aircraft destroyed between July and September 1940 were not due to enemy action and were not destroyed on operations.

German Aircraft Destroyed on Operations, May–September 1940

Type	Strength 29 June	Due to enemy action	Not due to enemy action	Total	Destroyed not on operations	Total destroyed	Losses as percentage of initial strength
Close recce	345	68	7	75	11	86	25%
Long-range recce	321	115	32	147	11	158	49%
Single-engine fighters	1,369	567	145	712	63	775	57%
Twin-engine fighters	367	304	25	329	16	345	94%
Bombers	1,758	862	180	1,042	100	1,142	65%
Dive-bombers	417	148	34	182	28	210	50%
Transport	531	191	19	210	18	228	43%
Coastal	241	58	45	103	17	120	50%
Total	*5,349*	*2,313*	*487*	*2,800*	*264*	*3,064*	*57%*

The attrition of aircrew led to a significant drop in the percentage of operations-ready crews present in the squadrons over the summer.

Percentage of Fully Operations-ready German Crews, July–September 1940

Type	7 Jul	16 Jul	23 Jul	30 Jul	3 Aug	10 Aug	17 Aug	24 Aug	31 Aug	7 Sep	14 Sep	21 Sep	Average
Bombers	75	73	69	70	67	71	61	66	65	66	65	59	67
Bf 110s	84	83	80	79	70	83	83	82	81	76	60	64	77
Bf 109s	81	81	83	82	78	85	81	76	74	74	81	77	79

Second, 'Ultra' decrypts began to help the British. These and recordings of PoW conversations helped steer British scientists to find methods of jamming the German radio signals that assisted the *Knickebein* night-bombing system; further assistance from 'Ultra' also helped nullify the next radio guidance systems: the *X-* and *Y-Geräte*.

By the time the German forces were ready for the attack on the Soviet Union, the Luftwaffe's strength was almost exactly the same as it had been at the start of *Fall Gelb*. It would go on to perform brilliantly in the opening days of *Barbarossa* but quickly found it could not provide cover over such a huge battlefield.

German Aircraft Strength

Type	11 May 1940	21 June 1941
Close recce	335	440
Long-range recce	322	393
Single-engine fighters	1,356	1,440
Nightfighters	0	263
Twin-engine fighters	354	188
Bombers	1,711	1,511
Dive-bombers	414	424
Ground-attack	50	0
Coastal	240	223
Total	*4,782*	*4,882*

Luftwaffe briefing with, from left to right: a *Leutnant* (pilot officer), an *Oberst* (group captain), ?, possibly a *Feldwebel* (flight sergeant). The *Kragenspiegel* colours are probably anti-aircraft arm of service (*hochrot* – bright red) although the *Oberst*'s could be *Karmesin* (carmine) denoting a *Generalstabsoffizier* (officer on the general staff). The map is of the Franco-German border along the Rhine, dividing Alsace (Elsass in German) from the state of Baden. Annexation by the Germans in 1940 – it actually became part of the Reich – meant that the population was liable for service in the German Army. 134,000 men and women from Alsace and Moselle (part of Lorraine) were drafted. They are known today as the 'Malgré-Nous/-Elles' (men/women, 'against our will') and some 32,000 (some say as many as 40,000) died.

Churchill is said to have called it the 'scourge of the Atlantic', and the Focke-Wulf Fw 200 Condor was certainly used to effect both as a bomber and as a maritime patrol aircraft. However, built as an airliner, with fewer than 300 airframes in total, it was too lightly built and there were never sufficient to fulfil both roles. The Fw 200 was also effective as a transport, delivering fuel to the Afrika Korps and supplies to surrounded German forces in Stalingrad and other Eastern Front pockets.

Opposite: A Luftwaffe bomber pilot posing in flight jacket and leather flying helmet. He's wearing the type K So/34 lightweight, tan-coloured combination summer flying suit – a step-in garment with zip and press-studs and tabs closure. The flying helmet is, strangely, the winter issue which was lined with fleece and designated the letter W for winter! It was crucial that the acoustic quality of the built-in microphone and earphones provided clear and audible transmission. The switch to activate the microphone could be on the headset or on a control panel and when activated connected the headset to the bomber's intercom system.

Above: The Messerschmitt Bf 110 was a twin-engined fighter-bomber that suffered high losses during the Battle of Britain while escorting slower bombers. The loss of its speed advantage and lack of manoeuvrability made it vulnerable to the nimble RAF fighters. Nevertheless, its long-range capacity allowed it to strike enemy shipping along the English Channel and the Mediterranean and Norway. The photograph shows Bf 110D-1 3U+CS of 8./ZG26 on escort patrol over the Mediterranean. The D-1s' two 1,200-litre external fuel pods had an impact on performance and were replaced with two 900-litre drop tanks for the Bf 110D-2. Note the direction-finder loop antenna under the fuselage. The Bf 110 later found success as a nightfighter using the FuG 202/220 Lichtenstein air intercept radar and, in its G-4 form, *Schräge Musik* upward-firing 20mm cannon.

Above: Designed in the early 1930s as a *Schnellbomber* (fast bomber), crew-friendly and affectionally known as the *Fliegender Bleistift* (flying pencil), the Do 17 was used extensively during the early war years but needed updating to increase its payload and range. Its successor was the Dornier Do 217. The Do 17 continued in various secondary roles, as a glider tug and trainer, for example, and many were transferred to other Axis air forces. The only surviving Do 17 was retrieved in 2013 from the Goodwin Sands in the English Channel, having been shot down in 1940. Behind it in the photo is a Junkers Ju 52/3m and in the far distance a Ju 87 Stuka.

Below: The Do 17 was also used as a reconnaissance aircraft. These Do 17Ps of 3(F)./Aufklärungsgruppe 123 (the F standing for *Fernaufklärung* = long-range reconnaissance) – the 3. Staffel emblem being a hammer and anvil – are seen at Fliegerhorst (Airfield) Langendiebach near Hanau in Hesse in early 1940. The unit provided reconnaissance for Luftflotte 3 over southern Poland, France and then in the Battle of Britain. The Do 17P was equipped with blind flying and landing instruments and the necessary camera equipment for reconnaissance work: either Rb 20/30 and Rb 50/30 or Rb 20/18 and Rb 50/18 cameras. Flying at 3,700m (12,000ft) meant that the four-man crew needed oxygen; planned 7,620m (25,000ft) altitudes were never achieved. The name *Flitzer*, next to the *Staffel* symbol, means 'speedster' but is also the word used by the Germans for streakers.

Originally designed under the guise of a civil airliner but for easy conversion into a bomber, the He 111P (above), produced in 1938, was the first of the type to have a stepless cockpit, meaning that the aircraft nose has no separate windscreen panels; essentially no breaks in the nose contour. This design helped make the aircraft more aerodynamic, thus improving speed and fuel efficiency although the inclusion of a nose-mounted gun caused a few problems. The He 111 saw action in the Spanish Civil War as part of the German Condor Legion. They were slow and not very manoeuvrable, poorly armed with a small payload. Nevertheless, they were the mainstay of the Luftwaffe's bombing campaign against Britain. Vulnerable to fighter attack even when given fighter escort, in the end their use was reduced to night activities. The He 111, however, proved a versatile aircraft, used as a tug for gliders, dropping a limited number of paratroops, sea-mine laying, reconnaissance and supply dropping (at Stalingrad, for example).

Above: Mirroring the many photos of RAF aircrew on standby in the hot summer of 1940, Hauptmann Karl-Heinz Greisert awaits the call to action on a forward airfield. 'Heino' Greisert had a brief but eventful life as a fighter pilot, starting in Spain with the Condor Legion, where he shot down his first enemy, a Polikarpov I-16. He served with II./JG 2 – Jagdgeschwader *Richthofen* – in France, adding six more kills, and took his tally to eleven during the Battle of Britain. In late November 1940 he became the temporary commander of the unit when the commanding officer went missing over the Channel. JG 2 was left in the west during *Barbarossa* and in February 1942 Greisert took part in the air defence of *Unternehmen Donnerkeil* – the Channel Dash by battleships *Scharnhorst* and *Gneisenau* from Brest to Germany – and also intercepted Lancasters heading towards Augsburg on 17 April 1942, shooting one down, his twenty-second victory. Transferred to the Eastern Front in May 1942 as Major and CO of III./JG 3, he died in combat with I-16s (the aircraft he had fought in Spain) with thirty-four victories to his name.

Below: During World War II photo-reconnaissance (PR) was an important method of intelligence gathering. Both sides used subterfuge to try to hide their airfields and troop concentrations from the overflying PR aircraft, and this included positioning wooden fakes that PR interpreters would see and misinterpret. The German word for one of these dummies is *Atrappe*. This photograph shows a good example of *Atrappen* –a Bf 109 and a Ju 88 – taken by a member of II./JG 53 based at Dinan who had helped set up the dummy airfield in June 1940. There's an amusing mention of this type of subterfuge in Schirer's 27 November 1940 diary entry. 'X tells me a funny one. He says the British intelligence in Holland is working fine. Both sides in this war have built a number of dummy airdromes and strewn them with wooden planes. X says the Germans recently completed a very large one near Amsterdam. They bed up more than a hundred dummy planes made of wood on the field and waited for the British to come over and bomb them. Next morning the British did come. They let loose with a lot of bombs. The bombs were made of wood.' The bombing part of the story is almost certainly apocryphal, although Pierre Antoine Courouble quotes Leutnant (later Oberstleutnant) Werner Thiel, who said he witnessed such an attack in October 1943 (see O'Malley, n.d.).

'Heino' Greisert (left) and a fellow pilot from II./JG 2 pose for the camera. Both men wear semi-official fleece-lined leather flying jackets, most made to the wearer's requirement and generally tailored to waist level. On the left Greisert wears leather 'Channel trousers' (*Kanalhose*), a term coined by aircrew involved in air activity over the English Channel. His boots are the standard-issue insulated-sole Luftwaffe flying boots. Unfortunately, after prolonged wear the nap of the sole flattened and any insulation was lost. They certainly weren't designed for long-distance walking. He holds a pair of gloves under his officer's flight cap (*Fliegermütze*) with its distinctive silver-aluminium piping. On his *Fliegerbluse* he has the ribbon of the EK II and just visible on his left breast is the EK I, pinned. Underneath that would be the pilot's badge. A collar and tie were standard wear. His colleague to the right wears standard wool-gaberdine breeches and high boots. On his head is a Luftwaffe officer's peaked cap and around his neck a silk scarf to prevent rubbing sores on his neck when searching the skies. The IK on the numberplate of the Mercedes Type 320 staff car identifies Schlesien or Silesia, an historical region of Poland. Note the photobomb character at the back.

Right, below and overleaf: This fabulous collection of Bf 109 images (and those of 'Heino' Greisert opposite and p. 222) is from the collection of Bruno Stolle, a fighter ace with thirty-five victories who ended the war a *Hauptmann* and also served in the postwar Luftwaffe. Here, he's an Oberleutnant wearing the EK II ribbon from his buttonhole and the EK I pinned onto his breast pocket. Below is his pilot's badge with an eagle clutching a swastika, an image of unbridled aggression and ferocity. His cuff-title – Jagdgeschwader Richthofen – was introduced on 25 March 1935. It was originally to be worn by the men of Fliegergruppe Döberitz, eventually renamed as Jagdgeschwader 2. In March 1943 Stolle was awarded the *Ritterkreuz*.

The iconic Junkers Ju 87 Stuka dive-bomber and ground-attack aircraft first saw combat during the Spanish Civil War. It was easily recognisable with inverted gull wings and non-retracting 'spatted' undercarriage (recalling the spats worn by prewar American gangsters). It was too slow to evade enemy fighters and when the Luftwaffe lost local air superiority – for example, during the Battle of Britain or in North Africa – it proved an easy target for Allied fighters. Nevertheless, it continued in service in many guises including as a tank-buster and a long-range anti-shipping bomber (Ju 87R, as here), until replaced by the Fw 190 ground-attack version. A dusty take-off (inset) could be hazardous especially if it impaired visibility, or more commonly it could clog engine systems, leading to engine failure, loss of power or other mechanical problems.

The Ju 88A-1 was a twin-engined multi-role combat plane, probably the Luftwaffe's most versatile aircraft. The original specification was for a *Schnellbomber* (fast bomber) able to carry a bomb load of 800–1,000kg (1,760–2,200lb) at speeds of 500kph (311mph). What was produced was an all-metal, stressed-skin, low-wing monoplane. The A-1 entered service in 1939. With a maximum speed of 450kph (280mph), it could carry a bomb load of 3,000kg (6,600lb), internally and externally. Protection was originally three 7.92mm MG 15s, at nose, rear cockpit and ventral gondola, but when this proved inadequate, more were added. Note the distinctive 'beetle's eye' nose glazing, the Bola undernose ventral defensive MG emplacement and annular radiators for engine cooling, set immediately in front of the engine and behind the propellers, that allowed for quicker cooling.

The Henschel Hs 126 was a tactical reconnaissance and army cooperation aircraft. TT1+CK of 2./Aufklärungsgruppe (H) 21 was photographed by a doctor in 33. Infanterie-Division in France in 1940. Its landing light is extended so the accident was probably a night-landing mishap.Slow and vulnerable, it was superseded by the Fieseler Storch. Hs 126s towed the DFS 230 gliders in the mission to free Mussolini on the Gran Sasso in September 1943.

Above: Bf 109E-3 of 8./III. JG 53 – the 8. Staffel of the third Gruppe of Jagdgeschwader 53 – near La Villiaze, Guernsey in July 1940. Note on the nose the red band which serves to hide the 'Pik As' – Ace of Spades – emblem. It was ordered as part of an official campaign to irritate and discomfort the *Geschwaderkommodore* (commander) Oberstleutnant Hans-Jürgen von Cramon-Taubadel who had fallen out of favour with Nazi officialdom because he'd married a woman with Jewish ancestry.

The Junkers Ju 86 was designed in the early 1930s. A medium bomber that was obsolete by 1939, it was remodelled as a high-altitude reconnaissance aircraft reaching heights in excess of 12,000m (40,000ft). Thought to be well out of the range of the Spitfire, the Ju 86P had the world's first pressurised cockpit. The British tried to combat this by modifying some Spitfires Mk Vc, reducing weight by 200kg (450lb) and replacing the armament of four .303 MGs with two 20mm cannon. Commanded by ex-Eagle Squadron Flight Lieutenant James C. Nelson (later Squadron Leader, AFC) who stayed with the RAF when the United States entered the war, this Special Spitfire Flight flew out of RAF Northolt and made interceptions at 12,800m (42,000ft) and then up to 13,400m (44,000ft), although the Spitfire failed to down the Ju 86P. But, showing they could be intercepted was enough to stop the missions.

Opposite below: Werner 'Vati' (Daddy) Mölders, here sporting a spotted silk neckerchief, was the first fighter pilot in aviation history to shoot down 100 aircraft. His early kills were with the Condor Legion, then he downed six over France, before being shot down and captured by the French. He was soon released when France surrendered, in time to fight over Britain and the Channel. By the end of the Battle of Britain his total was fifty-four. On the Eastern Front he commanded JG 51, the first Luftwaffe fighter group to claim 1,000 kills, and received the *Ritterkreuz des Eisernen Kreuzes mit Eichenlaub und Schwertern* (Knight's Cross with Oak Leaves and Swords), the second Luftwaffe pilot to receive the honour (Adolf Galland being the first). By the end of July 1941 Mölders had 115 kills to his name and had become too valuable a propaganda asset to risk in combat. Recalled to Germany, he became Inspecteur der Jagdflieger (Inspector of Fighters). He died on the way to Ernst Udet's funeral and was buried next to him; JG 51 adopted the name Mölders early in 1942.

9 Europe Under the Swastika

THE WORLD DIDN'T SEE exactly what happened in Poland after the surrender or it might have been a little more worried. *Kristallnacht* should have given it a good idea. The pogrom against German Jews on the night of 9–10 October 1938 saw thousands sent to concentration camps, thousands of places of worship destroyed and thousands of businesses looted. In Poland, an even more chilling occupation was planned that would result in over 6 million Polish deaths – over 20 per cent of the population – including 3 million Jews. This wasn't a pogrom; this was genocide.

The German intentions in the east – *Generalplan Ost* – intended to see the Poles and the Slavs killed off and the countries left free for settlement by Germans. Some Poles were identified as being capable of Germanisation; over 3 million others were used as slave labour. Following on behind the troops were the *Einsatzgruppen*, SS death squads that searched out for immediate execution the intelligentsia, the professional classes, the doctors and teachers and scholars – and, of course, the Jews.

Things were slightly different in the west. The immediate aftermath of the German conquest in the Netherlands, for example, was benevolent – the main reason being the need to keep the conquered areas locked down while fighting was still going on. Germans were supposed to pay for goods in cash; a suitable exchange rate was organised and many Dutch drivers were employed to move goods and troops. There was a curfew and all motor transport was immediately requisitioned by the German Army. Food was also taken, some to feed the troops, some to be sent back to the Reich.

After France fell, however, things began to change in the Low Countries. A *Reichskommissar* was appointed – Arthur Seyss-Inquart, who was hanged in 1946 after the Nuremberg Trials for overseeing a reign of terror in the country. Men and boys aged 16 or over were sent to Germany to work: the country provided over half a million labourers, half of whom were sent to Germany. Not every Dutchman was a forced labourer, however; over 25,000 were politically aligned enough with the Nazis to join the Waffen-SS. On the other hand, there was a significant resistance movement and many Jews were hidden from the Germans, the best known being Anne Frank, who survived for two years in hiding before being discovered. She and her sister died in early 1945, probably from typhus, in the Bergen-Belsen concentration camp, two of the more than 100,000 Dutch Jews who died during the war.

Food became scarcer as the war went on, as did medicine, and this reached a peak during the 'Hunger Winter' of 1944–45 when the Germans cut off food supplies to the west of the Netherlands after a general strike. Nearly 20,000 died, bringing the total Dutch wartime-related deaths, including Holocaust victims, to around 206,000. Other countries fared even worse famines: Greece suffered over 100,000 deaths from famine in 1941–42.

Similar events took place in every country under Nazi sway, with the German taskmasters appropriating not just food but also financial reserves, manufacturing resources, transport facilities and cultural artefacts. Military bases and training areas, airfields and ports were taken over for German use.

Manpower was the most obvious appropriation: in 1944 over a quarter of the German workforce was made up of slave labour, although some went under friendlier terminology as *Fremdarbeiter* (foreign workers), *Zivilarbeiter* (civilian workers) or *Militarinternierte* (military internees – PoWs). Most were *Ostarbeiter* (eastern workers) who were treated abominably, but not quite as badly as the concentration camp labourers, who died in their droves.

> In the summer of 1944, 7.6 million foreign workers found themselves employed in the Reich: 5.7 million civilian workers and a little less than 2 million prisoners of war. Of these, 2.8 million came from the Soviet Union, 1.7 million from Poland and 1.3 million from France, but there were people from almost twenty European countries deployed to work in the Reich. (Herbert, 2000)

There was resistance and there was collaboration – unsurprising when the circumstances are taken into account – but those countries who had been defeated by Blitzkrieg were to spend a terrifying four or five years under the swastika.

By the middle of 1940, the Germans had rearranged the borders of Europe. The first thing they did was to expand Germany itself, annexing areas: Luxembourg, elements of Poland (the main area, Generalgouvernement, was, like the Protectorate of Bohemia and Moravia, considered semiautonomous), Belgium, Alsace and Lorraine. Further territorial expansion would see other areas in Yugoslavia, Hungary and the Soviet Union annexed. Hitler's territorial plans extended into the west – Iceland and Greenland – south (northern Italy) and, of course, eastwards. This is, as the sign identifies, a 'Border crossing point! ID check. Exchange Reich money into Reich credit/cash notes.'

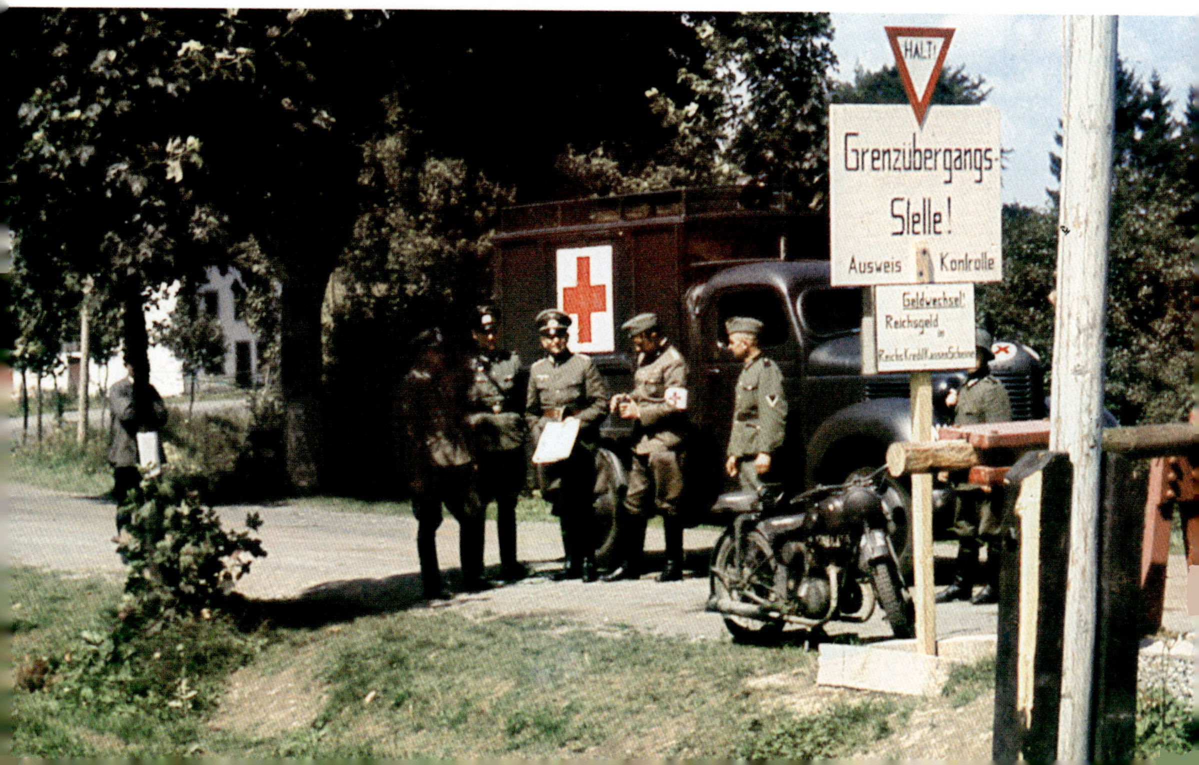

Salvage

One of the first things the Germans did was see what they could use in the conquered lands. The *Beutestäbe* were salvage staffs for captured (*Beute*) enemy materiel. There were a number of them for different types of equipment – including *Festungs-Pionier-Beutestäbe* – fortress salvage staff. Set up in August 1940, there were four *Panzerbeutestäbe*, who catalogued all the vehicles acquired by the Germans. Once they finished this job, the then General der Schnellen Truppen, Generalmajor Wilhelm Ritter von Thoma, reported back to OKW on nearly 3,500 captured French AFVs, of which around 1,000 were Renault R 35s and 1,200 were Hotchkiss H 35s or 39s. Most of these captured vehicles were employed by the Germans – sometimes for their original use as tanks, often as tractors or ammunition vehicles and sometimes as the chassis for *Panzerjäger* or SP guns, making use of captured weapons such as the Soviet 7.62mm anti-tank gun. Where possible, the Germans modified the vehicles to include a radio (many didn't have one), improved the commander's vision by adding a suitable cupola and added a crewmember to reduce the commander's tasks. The *Panzerbeutestäbe* were later absorbed into Panzer maintenance detachments deployed with the *Panzergruppen* on the Eastern Front.

The Germans made big use of *Beute* – looted – hardware: tanks, lorries, weapons. This Somua S35 is in Wehrmacht service near Lillehammer station, Norway in 1942. The S35s gave a good account of themselves in France and the Germans were able to take nearly 300 into service. They were used in secondary roles once they had been suitably modified. Note the German Fu 5 radio antenna on the right side of the superstructure above the four side-storage panniers, and the PzKpfw II cupola; most S35s received a PzKpfw III cupola.

About 3,000 Renault UE Chenillettes were captured in 1940. They were reused in various forms: as tractors, munitions vehicles, SP guns armed with the 3.7cm Pak 36 and a range of other vehicles.

360 Lorraine 37L tractors were captured and used by the Germans – as supply or ammunition vehicles, artillery observation vehicles, Marder I Panzerjäger and SP guns. Note the ammunition trailer.

Above:The National Socialist Workers' Party of Denmark (DNSAP) was founded on 16 November 1930 and welcomed the Germans into their country. It wasn't very popular. Even in the 1943 election it could only take 2.1 per cent of the votes. This is a local DNSAP rally in 1940. The occupation continued until *Befrielsen* (the Liberation) in May 1945 by British forces – except for the island of Bornholm, which was taken by the Soviet Union.

Opposite above: Two men heading to a forced labour camp. A project with the participation of the St Pölten University of Applied Sciences identified 2,115 Nazi camp sites in Austria, 372 of them in Lower Austria. They included Mauthausen concentration camp, whose subcamps were built near armament factories. An example of these was Kematen Forced Labour Camp, which was planned to construct tunnels suitable for production of Messerschmitt Me 262s. There were also PoW camps such as Stalag XVII B in Krems-Gneixendorf, one of the largest built, where at times more than 60,000 prisoners of war were detained.

Opposite below: Two of the occupation forces in Denmark in front of their Opel Olympia. Until 1943 there was an uneasy arrangement of government in Denmark under the administration of Reichsbevollmächtigter (Plenipotentiary) Cécil von Renthe-Fink. After that the country was placed under direct military occupation, leading to the rescue of the Jews by sending them to Sweden. The main dependencies of Denmark were Iceland and Greenland. The former was occupied by Britain on 10 May 1940 before being handed over to the USA, which was then neutral. It became an independent republic on 17 June 1944. The Danish envoy to the United States signed an agreement with the USA, authorising it to defend Greenland.

Above: There were five *Soldatenheime* in central Paris, an *Offiziersheim* and three *Fliegerheime*. Here, soldiers could lodge, relax, eat and socialise. While most soldiers went home on leave, some would spend time in such entertainment centres as Paris. There were guidebooks that provided information on German military facilities and the *Soldatenheime* were ideal places to base themselves for tourism and relaxation.

Opposite: The classic view of occupation: German soldiers – these are *Kriegsberichter* – and the Eiffel Tower. Hitler visited on 24 June. Paris became a centre for soldiers' rest and relaxation: everything from *Soldatenkaffee Madeleine* (coffee shop), three *Soldatenkinos* (cinemas), three *Soldatentheater*, a *Soldatenkaufhaus* (department store) to a range of vetted brothels and their nearby 'Responsible remediation agencies'. Brothel visitors were informed that 'Visitor cards issued in the houses must be retained for at least five weeks for future reference.'

On 26 July 1940 the Hôpital Lariboisière became a *Kriegslazarett*, a hospital for the Wehrmacht. The doctors, nurses and, of course, the patients were German.

BOISIERE

A military signpost in Paris. At the top, Der Militärbefehlshaber in Frankreich (the Military Commander in France) was based at the Hôtel Majestic. Paris endured a hard four years with the German administration clustered in the centre. Der Kommandant von Gross-Paris (the Commandant of Greater Paris) was at the Hôtel Le Meurice in rue de Rivoli (the best-known holder of the position being General Dietrich von Choltitz, who didn't follow orders in 1944 and didn't leave the city destroyed). The Gestapo was at 11 rue des Saussaies, the French version (the Carlingue) was at 93 rue Lauriston and the Sicherheitsdienst, the intelligence agency of the SS, at 84 Avenue Foch.

A large number of the male French population – 1.5 million PoWs – went to Germany, where they were used as labourers. In 1942 and 1943 this led to Vichy's two compulsory labour laws, the second setting up the Service du Travail Obligatoire (Compulsory Work Service) that saw 650,000 workers sent to the Reich. But as this poster says, 'Every hour of work in Germany is a stone added to the rampart that protects France.'

Magdeburg

Deutsche
Reichsbahn

Railways played a crucial role in Blitzkrieg, moving men and matériel to jump-off points, carrying supplies and reinforcements. After the fighting, the Germans took over French Railways – SNCF, the Société Nationale des Chemins de Fer Français – and used its lines to ship war booty to Germany: everything from food and manufactured goods to raw materials and military equipment. Of course, it was also used to deport Jews and others to the death camps. German wagon classes were given the names of German cities as exemplified here: the Karlsruhe class (inset, seen at Magdeburg main station) equipped with a brakeman's box; the München and Kassel classes (main photo) which were covered goods wagons.

Above: Erich Fellgiebel, a signals specialist who had been head of the cipher bureau in the Ministry of the Reichswehr, on 1 March 1938 became Chef des Wehrmachtnachrichten-Verbindungswesen (Chief of the Wehrmacht Intelligence Liaison Service). On 1 August 1940 he became General der Nachrichtentruppe (Head of the Intelligence Corps), and he's seen inspecting a temporary airfield in France. Note the red signals flash on the windscreen. Fellgiebel was hanged on 4 September 1944 as an important member of the 20 July Bomb Plot.

Opposite: Uniform maintenance in the sunshine – the army used France for training and resuscitating units that had been mauled elsewhere. They charged the French for this: over 700 billion francs for the four years of occupation. The rip-off didn't end there: the overvalued exchange rate meant that the Reichsmark-wielding soldiery could buy any consumer goods that they wanted.

Below: To the victors the spoils – in this case French and British helmets allow a bit of hamming for the camera.

Mont-Saint-Michel was as much of a tourist trap then as it is today. It's said that some 325,000 Germans made the trip as this *Fallschirmjäger* has done. Things weren't as easy for the French people. The Germans requisitioned 21 per cent of the meat, 13 per cent of the wheat and large amounts of milk, butter and other foodstuffs. By the time occupation was over, on average 71 per cent of the French household budget went on food.

Inset: The fighting's over and what better way to enjoy the sunshine for these 4. Panzer-Division soldiers? The somnolent chap is keeping his bayonet close and wears a *SA-Wehrabzeichen*. This award was originally instituted by Ernst Röhm, the *Stabschef* of the *Sturmabteilung* who died in the 'Night of the Long Knives' along with at least 200 other political enemies, as the *SA-Sportabzeichen* (SA Sports Badge). Hitler changed its name to *SA-Wehrabzeichen* (SA-Defence Badge) on 19 January 1939. It was one of the few political badges allowed to be worn on uniforms. The *Seitengewehr* SG 84/98 für Karabiner K98 (bayonet for the Kar98k) and its *Seitengewehrtasche* (scabbard) came – depending on date – with wooden or Bakelite handles, screwed or rivetted together. Collectors look for identical numbers on bayonet and sheath as they were issued together. It was prohibited to sharpen the bayonet.

The defence of occupied Europe wasn't something considered highly by the Germans in 1940. There was little likelihood that an enemy driven off so recently would try to come back. It wasn't until Führer Directive No. 40 of 23 March 1942 that defensive measures were specified: 'In the near future the European coasts will be exposed very seriously to the danger of enemy landings.' This was prescient. On 19 August 1942, the Allies raided Dieppe.

Inset above: Military installations received a modicum of protection – especially on the Channel coast from where *Unternehmen Seelöwe*, the invasion of Britain, would take place. This is a leFH 18 under camouflage netting.

Inset below: The Normandy beaches increased in importance over the years as an invasion became more likely. Slowly, the Atlantic Wall grew in size and depth, built by Organisation Todt and, later, slave labour as Hitler directed. By 1944, minefields, beach defences and a range of pillboxes created an impressive barrier that took a heavy toll on the Allied forces in June 1944.

Above: German radar played an important role in air defence of the German-held continent, especially when used to vector nightfighters onto the bombers. The first German defences were set up in 1940 and were given the name of the man who established them, Oberst Josef Kammhuber. The Kammhuber Line used two radars in unison: the longer-range less-accurate Freya (square, upright) and the accurate close-range Würzburg (small, round, on wheels). The British were able to develop countermeasures thanks to Operation *Biting*, when a commando raid on Bruneval in February 1942 acquired elements of the radar. This led to the development of 'Chaff' and 'Window' white noise radar jammers. Photograph shows the radar station at Douvres-la-Délivrande in Normandy later in the war.

Below: Acquiring access to the French coast and its ports and harbours gave Germany a huge boost in the war against Britain and opportunities to bust the blockade. The key element was housing the U-Bootwaffe: it had almost brought Britain to its knees in World War I and came very close to doing so again in World War II. To this end, the Kriegsmarine commissioned into service some 1,150 or so U-boats during the war and they intercepted over 650 convoys and sank around 3,000 Allied ships. Vizeadmiral Karl Dönitz officially opened the first of the Saint-Nazaire pens – there would be fourteen in total – on 30 June 1941. This Type VIIC is about to leave Saint-Nazaire in 1942.

In the background, Napoleon's Pont de Pierre crossing the Garonne in central Bordeaux; in the foreground, three Kriegsmarine M35-class minesweepers. They are flying the Wehrmacht battle flag and carry the main tools of the minesweeping trade: the paravanes (white bomb-shaped objects) that were trawled from the vessel to snag the cables of tethered mines and cut them. Once they drifted to the surface they could be detonated. Note, too, the closed panels either side of the turreted 10.5cm gun: they hide the gun optics and would be opened in combat. The central vessel has the 'goalpost' style davit for its launch, the other two the more traditional style. Bordeaux was also the home of BETASOM, the Italian submarines that fought alongside the Kriegsmarine in the Atlantic.

Inset: The threat of Allied bombers led to the construction of huge U-boat pens and manufacturing facilities. This is La Pallice, La Rochelle's port, where construction of U-boat bunkers began in April 1941. The first of ten was completed by October of that year.

Above: French ships at anchor in Marseille, 1940. The French had a significant fleet and it was one the British wanted to neutralise. After the French armistice, on 3 July 1940, at Mers-el-Kébir in French Algeria and later on 8 July off Dakar, the Royal Navy bombarded the French fleet and successfully damaged a number of its capital ships. There were some engagements between French ships under the Vichy government and the Royal Navy, but nothing major, and on 27 November 1942, following the German invasion of Vichy France, the French scuttled the major part of their fleet – nearly eighty ships including three battleships and seven cruisers – in Toulon harbour rather than have them fall into German hands.

Opposite above: Hot food from a *Feldküche*. Note the men are in belt order, which meant that all the weight of their equipment was on their hips – uncomfortable and impractical and the reason why soldiers wear shoulder straps from which they hang their equipment. There were a considerable number of German troops in Norway during the war. The northern end of the Atlantic Wall, it had several major Kriegsmarine and Luftwaffe bases that played an important role in the attacks on the Allied Arctic convoys to Murmansk.

Opposite below: The main components of the German Flak units in Norway in 1940 were from Flak-Regimenter 32, 33 and 141. They took up positions around Bergen, Kristiansand, Narvik, Oslo and Stavanger. Later, at the end of April 1941, a Flak school opened near Oslo. The photograph shows a local child talking to the guncrew of a 2cm Flak 30/38. Note the ammunition container next to the youngest member of the crew.

Above: As elsewhere in conquered Europe, on the surface life returned to some sort of normality in Norway after the initial fighting. However, Norway was under a heavy-handed occupation with many troops, a strong naval presence and the occupation lasted till the very end of the war, 8 May 1945. The Reichskommissariat Norwegen was under Josef Terboven, who was a hardline Nazi and was disliked enormously by the Norwegians. He was responsible for setting up concentration camps, persecuting and deporting Jews, and summary execution of hundreds of civilians in his attempts to supress the resistance movement. His scorched-earth policy in the closing months of the war led to significant destruction in the north of the country. This is a ferry at Tromso in 1940.

Opposite: Trondheim – Drontheim in German – was the largest Kriegsmarine base in Northern Europe, the HQ of 13. Unterseebootsflottille. The Nazis had great plans for Norway and intended to build a cultural hub of the north, New Trondheim, promoting their white supremacist concept of 'Aryan' racial superiority. They also planned to expand the naval base and in 1941 began the construction of two U-boat bunkers, only one of which – Dora I – was completed by 1945. U-boats made major use of Norwegian harbours. There were 275 patrols out of Bergen, 168 out of Narvik and 142 out of Trondheim during the war. The photograph shows a U-boat in Trondheim Fjord.

Coastal defence of Norway and Denmark was treated very seriously by the Germans. Initially the defences were primarily AA and artillery units in open platforms. Later in the war huge numbers of concrete bunkers were built as the Atlantic Wall took shape. At Trondenes, the big 40.6cm (16-inch) guns of Batterie Dietl had a range of 50km (30 miles). They had initially been intended for battleships *Friedrich der Grosse* and *Grossdeutschland* that were originally scheduled for completion in 1944. This is a Canon de 155 Grande Puissance Filloux (GPF), in German hands, the 15.5cm K 418(f). The French had 450 in service in 1940 and the Germans had at least fifty still in service for coastal defence in 1944.

Above: No sooner had the fighting finished than the German authorities started their persecutions, deportations and campaign of murder. The Warsaw Ghetto was set up in November 1940. Walled off from the rest of the city, its population reached almost 450,000 to start with. It was razed in May 1943 on Hitler's orders following the Ghetto Uprising, its population dead or deported to the Treblinka death camp 80km (50 miles) north-east of the city.

Opposite: This *Obergefreiter* on guard in Poland 1940 is wearing two sports badges: the top one is the SA Sports Badge in Bronze; the lower that awarded by the DRL – Deutscher Reichsbund für Leibesübungen (German Imperial Commission for Physical Exercise). The impending Summer Olympics of 1936 led to a great national enthusiasm for sport and the Nazis took advantage of the propaganda. In 1938, the DRL became the NSRL, the Nationalsozialistischer Reichsbund für Leibesübungen (National Socialist League of the Reich for Physical Education) and was placed under Nazi Party control.

Once the fighting had finished in Poland – and the other conquered nations – the military occupation authorities moved in to pacify as necessary and govern. There was a net of headquarters at various levels with both administrative and military police components and such other units assigned as needed. Often, it was these units that were involved in the nitty-gritty of the Holocaust.

As many as 1.8 million French servicemen became PoWs in 1940: most went to Germany and ended up being put to work. Their mortality rate wasn't as severe as the PoWs of other nations – particularly the Soviets – but they were badly treated when they got home at the end of the war because they had surrendered rather than dying for France. It took until the 1950s for many to receive veterans' benefits.

Above: The treatment of white French PoWs was vastly better than that of their colonial troops. As many as 3,000 of the 120,000 colonial troops who surrendered were executed immediately. Subjecting them to abuse, torture and execution, the Nazis treated them as they would do the Soviets in 1941. Nazi racial theories classed Black people as subhumans, and while most of the massacres in France were committed by the heavily indoctrinated SS, there's no doubt that this sort of summary execution – and how the authorities turned a blind eye to it – contributed to the German Army's descent into dehumanisation and war crimes that would characterise its conduct of the war in the east.

Below: A French girl talks to her father through the bars of a German prison. Repression began as soon as the Germans entered France. On 4 June, Auguste Gras became the first of some 2,500 sentenced to death by Wehrmacht military courts. Gras was shot on the 5th in the Aisne *département*. On 4 July it was made official that German penal law became applicable in the occupied areas of France. One article of the German Military Code said, 'any act harmful to the security of the occupying troops, as well as any violation of a decree from the *Führer* or his representative, [was] a reprehensible act which [had to] be punished just as though it had been committed on the territory of the *Reich*.' (Eismann, quoted in Fontaine, 2007). That same day, Etienne Achavanne was shot in Rouen for cutting a telephone wire. It wasn't just the Germans who were involved in the repression. On 3 September the Vichy government adopted the Statut des Juifs (Statute on Jews). It would lead to round-ups and deportations. Communists were also targeted.

Conclusion – Ready for *Unternehmen Barbarossa*

By autumn 1940, the dust had settled on Germany's blistering resort to arms and the whole of western Europe had succumbed, save for Britain. Soon, the German Army would move against its existential foe, the communist Soviet Union. The reasons for its success were much discussed and the snappily titled new approach to warfare, Blitzkrieg, was heralded as the reason the balance of power shifted so dramatically against France, which before the war had been seen as the strongest military power in Europe.

Both sides had similar numbers of weapons and many of the Allies' were better than the Germans'. The Luftwaffe made a big difference, as did the brilliant plan of campaign and the leadership qualities of the junior officers and NCOs, whose adherence to *Auftragstaktik* allowed flexibility in attack and the ability to suit the method of attack to the situation on the ground. Perhaps even more important was the fact that the Germans made great use of radios – particularly to control their Panzers. The use of gliders and *Fallschirmjäger* also provided an unexpected element of surprise to the initial attacks.

But in amongst all the success, it's worth looking back on the events at Hannut and Gembloux between 12 and 15 May, where French 1re Armée held the German attacks and inflicted a significant reverse on Heeresgruppe B. Although the French were forced to retreat because of Heeresgruppe A's brilliant thrust to the Channel, nevertheless the Germans had to fight their way through the French at Lille while the British retreated to Dunkirk. The bulk of the BEF and a large part of 1re Armée escaped. As Gunsburg put it: 'The German victory proved incomplete; the war did not end in the summer of 1940; and the triumph of the Blitzkrieg proved short-lived. The Allied combatants of Gembloux can justly claim a part of that final vindication for themselves.'

US military eyewitnesses of the fighting provided their thoughts back to the US Army in 'Outstanding Impressions of the German Army. (Based on observation of operations in Poland and Holland.)':

A Aggressiveness.

The entire military establishment is conceived, organised, equipped and trained for offensive action.

B Extreme Coordination and Balance.

1. Tactical groups are organised and equipped for the task in hand. They are given the means to accomplish the mission, and they are self-contained.
2. Complete coordination of elements is accomplished by combined training, detailed rehearsal of important operations, and adequate mechanical liaison, such as radio, planes, etc.

C. Superiority of Force, Both in Manpower and Materiel

1. A superior air force is the most decisive single factor in German success. The Germans have amplified the capabilities of mechanized ground forces.
2. Cost in men and materiel is disregarded when vital objective is to be taken. Quantity and speed of production, rather than quality and long life, have governed in the manufacture of materiel.

D. Thorough Preliminary Internal Preparation (Fifth Column)

E. Complete Information on Capabilities of Opposing Countries

Adequate military intelligence is available. Detailed and minute knowledge is made possible by the fifth column.

These factors played a pivotal role in the west and would do so again in the east in 1941, but before *Barbarossa* the Germans were drawn into conflict in the Balkans. The Italians had invaded Greece in October 1940 and had been beaten back by a strong counterattack. Britain had an obligation to support Greece and provided RAF squadrons and troops in Crete. Hitler was concerned that the British might bomb the Romanian oilfields – fuel was always a critical logistical problem for the Germans – and was also worried about Mussolini's standing in his own country. He resolved to come to his ally's assistance and in April 1941 launched *Unternehmen Marita.*

It was as successful as all the other attacks had been. Greek and British defences were blown away and two months later the Germans were masters of the Balkans, Greece and Crete – although the aerial assault on the island saw so many casualties and came so close to failure that the Germans would never again authorise a major *Fallschirmjäger* aerial assault. This probably saved Malta and because of that, it could be said, the British in North Africa.

However, the Mediterranean became a sideshow on 22 June 1941 when the Germans launched *Barbarossa* and the last great Blitzkrieg attack saw the fate of the Soviet Union – and with it any hope of freeing western Europe – hanging in the balance.

A mound of French equipment left behind by soldiers starting five years of servitude to their conquerors.

Inset: Britain fought on—supported by the British Empire, in particular Canada who immediately supplied troops to help defend against the anticipated invasion. While the Battle of Britain raged, British forces did what they could to pre-empt a German invasion fleet. The only way to take the fight to the enemy at this stage of the war was in the air. British Intelligence suggested the Dortmund-Ems Canal was being used to ferry barges needed for the invasion. A choke point was north of Münster in Gelmer. The Münster aqueduct raids ensued—there were 15 of them in July and August 1940. The final raid was made on 12/13 August by five Hampden bombers of Nos 49 and 83 Squadrons. Heavy Flak defences saw all the attacking aircraft damages, two of them shot down and the pilot of the final Hampden, Flt Lt Roderick Learoyd, was awarded the Victoria Cross. This photograph shows camouflage covering repaired bridges over the canal.

Below: Amongst the booty the Germans picked up in France, the SS *De Grasse* (right) had had a chequered career. Started in 1921 at Cammell Laird and finished in France because of strikes, it was launched in 1924 and began its North Atlantic service with a crossing from Le Havre to New York. Used as a troopship when war began, it was in Bordeaux when captured by the Germans. It was used as an accommodation ship for German troops at Blaye on the Gironde before being handed over in 1941 to the Italians, who used it as a depot ship for BETASOM. Given to the Vichy French in 1942 and used as a training ship, it was scuttled again by the retreating Germans in 1944. Refloated postwar and rebuilt with a single funnel, it became *Empress of Australia* in the coronation year of 1953. Subsequently renamed *Venezuela*, it ran aground in 1962 and went to the breaker's yard.

The troops return home. Flowers, flags and fervour greeted them as they returned to Germany after the fighting. It really had been a lightning war and now it was all over. Versailles had been avenged. Hitler had proved himself a political and military giant. And now for the years of peace …

On 6 April 1941, the Wehrmacht continued its unstoppable advance, this time in Greece. Forced to come to their Italian ally's assistance following a failed attack in late 1940, once again the Germans mixed daring pinpoint airborne operations with the now tried-and-tested combined-arms assault. By 30 June 1941 it was all over: continental Greece had fallen. The island of Crete had fallen at the end of May – but at great cost. It was the last significant German airborne operation, and the two sides took very different views on the battle. The Allies saw just how sizeable airborne operations could influence the course of a battle and went on to prepare troops for major operations. On the other hand, the butcher's bill was too high for Hitler. He consigned the *Fallschirmjäger* to smaller battles and most spent the rest of the war as elite infantry. This is a file of *Gebirgsjäger* – General Franz Böhme's XVIII. Gebirgskorps had attacked the mountainous Metaxas Line along the Greece–Bulgaria border.

Crossing the River Save in Yugoslavia. Note the MP's traffic control paddle and the tactical markings on the light off-road passenger car: the yellow marking identifying 14. Panzer-Division, the white a motorised infantry unit. Built by Stoewer, BMW and Hanomag, it wasn't as good as the VW Kübelwagen, but over 15,000 were built.

A knocked-out British Cruiser Tank A10 Mk Ia of C Squadron, 3rd Royal Tank Regiment, 2nd Armoured Division. 1st Armoured Brigade was detached from the division in North Africa and assigned to the expeditionary force to Greece. It arrived with two armoured regiments (Cruisers of 3rd Royal Tank Regiment and light tanks of 3rd Hussars) in early March 1941 and on 9 April it came under command of the 6th Australian Division. Between 10 and 18 April, it was involved in the major battle at Mount Olympus. It then fell back and was evacuated back to Egypt on 29 April 1941.

After Greece had fallen, next stop for the Panzers was the Soviet Union. At the pinnacle of its power, Hitler and his generals thought the German Army would be able to defeat the inferior Red Army quickly – look how the small Finnish nation had held them at bay during the Winter War. *Unternehmen Barbarossa* began on 21 June 1941 with the final Blitzkrieg of the war.

It took four years before the effects of the 1940 Blitzkrieg were expunged. On 26 August 1944, Paris celebrated liberation, with De Gaulle striding down the Champs-Élysées followed by General Philippe Leclerc's 2nd Armored Division – its Black African troops having been substituted by whites from other divisions so as not to upset the segregated US Army. The British – whose Commonwealth forces were unsegregated – put up no resistance to this racism.

The American proponent of Blitzkrieg: General George S. Patton's Third (US) Army had a close relationship with the support aircraft of XIX Tactical Air Command. Brilliant tank commanders like Creighton Abrams (after whom the US Army's latest main battle tank is named), 4th Armored Division's Major General John S. Wood and 6th's Major General Robert W. Grow saw the Third Army exploit the gap opened up by Operation Cobra and blitz through France to the German border. The key elements of blitzkrieg – speed, air support and the concentration of tanks into an armoured spearhead – were used to great effect after Patton's army bounced the Rhine and surged through Germany and Austria, ending the war in Czechoslovakia. Here, Patton is seen on his visit to Paris in early August 1945.

Bibliography

Printed Books

Buchner, Alex. *The German Infantry Handbook*. Schiffer Publishing, 1991.

Campaign Study No 2: The German Campaign in Norway. War Department, 1942.

Chamberlain, Peter & Doyle, Hilary L. *Encyclopedia of German Tanks of World War Two*. Arms & Armour Press, 1978.

Clark, Lloyd. *Blitzkrieg: Myth, Reality and Hitler's Lightning War – France, 1940*. Atlantic Books, 2017.

Davis, Brian L. *Uniforms and Insignia of the Luftwaffe Volume 1: 1933–1940*. Arms & Armour Press, 1991.

Davis, Brian L. *Uniforms and Insignia of the Luftwaffe Volume 2: 1940–1945*. Arms & Armour Press, 1991.

Davis, Brian L. *German Army Uniforms and Insignia 1933–1945*. Brockhampton Press, 1998.

Forty, George & Duncan, John. *The Fall of France*. Nutshell, 1990.

Guderian, Generaloberst a.D. Heinz. *OKH Project #7: Representation of Armored Interests 1938–1945*. USAREUR Historical Division, 1952.

Hart, Stephen & Hart, Russell. *Blitzkrieg: The Invasion of Poland to the Fall of France*. Osprey, 2021.

Information Bulletin No. 54: West Front Operations May 1940. War Department, August 1940.

Jentz, Thomas L. *Panzer Truppen 1*. Schiffer Military History, 1996.

Kennedy, Major Robert M. *The German Campaign in Poland (1939)*. Department of the Army Pamphlet No. 20-255, 1956.

Maiwald, Barbara. *Feldküche und Co*. Motorbuch Verlag, 2018.

Manstein, Erich von. *Lost Victories*. Presidio Press, 1994.

Murray, Williamson. *Strategy for Defeat: The Luftwaffe 1933–1945*. Airpower Research Institute, 1983.

Office of Strategic Services (OSS). *Führer Directives and Other Top-Level Directives of the German Armed Forces 1939–1941*. OSS, 1948.

Pallud, Jean Paul. *Blitzkrieg in the West: Then and Now*. After the Battle, 1991.

Sáiz, Augustín. *Deutsche Soldaten*. Casemate, 2008.

Schirer, William L. *Berlin Diary, 1934–1941: The Rise of the Third Reich*. PRC, 1997.

Special Bulletins 9 and 26: Attempt to Capture The Hague by the 22d German Division. War Department, 1940.

Special Bulletin No 32: Account of Operations Between May 10 and the Armistice at Compiegne. War Department, March 1941.

Zaloga, Steven J. & Rodríguez, Felipe. *New Vanguard 327 German Tanks in France 1940.* Osprey, 2024.

Ziemke, Earl F. *The German Northern Theater of Operations 1940–1945.* Department of the Army Pamphlet No. 20-271, 1959.

Articles and Online Material

Alexander, Martin S. 'After Dunkirk: The French Army's Performance against "Case Red", 25 May to 25 June 1940', *War in History*, Vol. 14, Issue 2, April 2007.

Doughty, Robert A. 'The Myth of Blitzkrieg', in *Challenging the United States Symmetrically and Asymmetrically: Can America Be Defeated?* Strategic Studies Institute, US Army War College, 1998.

Fontaine, T. 'Chronology of Repression and Persecution in Occupied France, 1940–44', from www.sciencespo.fr/mass-violence-war-massacre-resistance/en/document/chronology-repression-and-persecution-occupied-france-1940-44.html, 2006.

Gunsburg, Jeffery A. 'The Battle of Gembloux, 14–15 May 1940: The "Blitzkrieg" Checked', *Journal of Military History*, Vol. 64, Issue 1, January 2000.

Harris, J. P. 'The Myth of Blitzkrieg', *War in History*, Vol. 2, Issue 3, November 1995.

Herbert, Ulrich. 'Forced Laborers in the Third Reich: An Overview', *International Labor and Working-Class History*, Vol. 58, Fall 2000.

Hitler, Adolf, *Mein Kampf.* Trans. James Murphy. Project Gutenberg Australia, ebook No. 0200601h, 2016.

Jersak, Tobias. 'Blitzkrieg Revisited: A New Look at Nazi War and Extermination Planning', *Historical Journal*, Vol. 43, Issue 2, June 2000.

Liddell Hart, Capain B. H. 'How Hitler Broke Through in the West', *Military Review*, March 1957.

McGrath, John J. 'A Motorized Infantry Regiment Crosses the Meuse River, May 1940', in Wright, Donald P. (ed.). *16 Cases of Mission Command.* Combined Studies Institute Press, 2013.

Murray, Williamson. 'The German Response to Victory in Poland: A Case Study in Professionalism', *Armed Forces & Society*, Vol. 7, Issue 2, Winter 1981.

O'Malley, David. 'Wood for Wood', article about Pierre Antoine Courouble and *Atrappen.* www.vintagewings.ca/stories/wood-for-wood?utm_source=chatgpt.com, n.d.

Shoemaker, Lt Col. John O. 'Sichelschnitt – Evolution of an Operation Plan', *Military Review*, March 1962.

Widder, Major General Werner. 'German Army, Auftragstaktik and Innere Führung: Trademarks of German Leadership', *Military Review*, September–October 2002.

Photo credits

John Avis: 88.
Matthieu Comas: 181TL&TR.
Michael Sobotta: 7T, 11T.
Ian Stewart Spring/pixpast.com: 7B, 8T&B, 10T&B, 11B, 12T&B, 13, 16, 17TR, 17B, 19, 20T&B, 23, 29B, 32–3, 37T&B, 42–3, 47 inset, 52B, 53, 54, 55, 56, 56–7, 58–9 and inset, 62, 63T, 64–5 and inset, 66–7 and inset, 68, 69, 70–1, 72–3, 75, 76T, 78–9, 80, 82–3, 86–7, 89T, 92–3, 102 inset, 104–5, 107 inset, 110–1, 112–3, 114–5, 116–7, 118, 119, 120–1, 122–3, 128–9 and inset, 130–1, 132–3, 134–5, 136–7, 138–9, 139B, 140–1, 142–3, 145T&B, 146–7, 148–9, 150–1, 153B, 154–5, 156, 160–1, 168T&B, 169, 171T&B, 172–3 and inset, 176T, 176–7, 178B, 180, 181B, 182T&B, 183T&B, 184T&B, 185T&B, 186, 187TR&B, 190B, 191, 192, 193T&B, 195, 196T&B, 197, 198–9, 202T&B, 205B, 206B, 210T, 210–1, 215T, 216, 217, 219B, 222–3, 232–3, 235, 237, 238, 239T, 240T&B, 241, 243, 244–4 and inset, 246, 247, 248–9 and inset, 250, 251T&B, 252–3 and inset, 255 both insets, 256T, 256–7, 258–9 and inset, 260, 261T&B, 262, 263, 264–5, 266, 268–9 and inset, 270–1, 271B, 273, 274–5 and inset, 277, 280–1, 282–3, 284T&B, 285, 288.
Akira Takiguchi: 17TL, 24–5, 27, 29T, 30, 38–9, 40T, 40–1, 46–7, 48–9, 50–1, 52T, 63B, 76–7, 81, 84–5, 89B, 90–1, 96, 97, 98–9, 100–1, 106–7, 152, 155 inset, 162 inset, 178–9, 188–9, 198T, 200T, 200–201, 203T, 204, 215B, 222T, 224, 225T&B, 226–7 and both insets, 229 inset, 234B, 239B, 242, 267, 278–9.
David Williams: 31T&B, 33T, 34–5 and inset, 36, 44–5, 51 inset, 102–3, 144, 153T, 157, 158–9 and inset, 161 inset, 162–3, 164–5 and inset, 166–7, 170, 174, 175T&B, 187TL, 190T, 203B, 205T, 206T, 207, 208, 209T&B, 218–9, 220T, 220–1, 228–9, 230–1, 234T, 254–5, 276.

Captured Renault R35s of 4e Division Cuirassée (armoured division). Armed with a short 37mm SA18 main gun and coaxial 7.5mm MAC31 MG, unusually for French armour of the time it also had a small pintle mounting for an AA gun at the rear of the turret. Its main drawback was that very few had a radio fitted. Note the red club turret marking. This indicates a vehicle of the 4th Platoon (clubs) of the 3rd Company (red). 1st Company was identified in blue and 2nd white. 1st Platoon were spades; 2nd, hearts; 3rd, diamonds